Reflected in Water

By the same author
Anarchy in Action
Streetwork: The Exploding School (with Anthony Fyson)
Vandalism
Tenants Take Over
Housing: An Anarchist Approach
The Child in the City
Art and the Built Environment (with Eileen Adams)
Arcadia for All (with Dennis Hardy)
When We Build Again
Goodnight Campers (with Dennis Hardy)
Chartres: The Making of a Miracle
The Allotment: Its Landscape and Culture (with David Crouch)
The Child in the Country
Undermining the Central Line (with Ruth Rendell)
Welcome, Thinner City: Urban Survival in the 1990s
Images of Childhood (with Tim Ward)
Talking Houses
Freedom To Go: After the Motor Age
Influences: Voices of Creative Dissent
New Town, Home Town
Talking Schools
Talking to Architects

Reflected in Water

A Crisis of Social Responsibility

Colin Ward

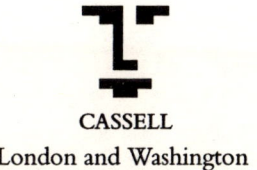

CASSELL
London and Washington

Cassell
Wellington House
125 Strand
London WC2R 0BB

PO Box 605
Herndon
VA 20172

© Colin Ward 1997

All rights reserved. No part of this publication may be reproduced or transmitted in any form or by any means, electronic or mechanical, including photocopying, recording or any storage information or retrieval system, without prior permission in writing from the publishers.

First published 1997

British Library Cataloguing-in-Publication Data
A catalogue record for this book is available from the British Library.

Library of Congress Cataloging-in-Publication Data
Ward, Colin.
    Reflected in water : a crisis of social responsibility / Colin Ward
      p. cm. (Global issues)
    Includes index.
    ISBN 0-304-33567-3 (hardcover). ISBN 0-304-33568-1 (pbk.)
      1. Water resources development. 2. Water rights. 3. Water-supply—Social aspects. 4. Water-supply—Political aspects.
    I. Title. II. Series
    HD1691.W317 1996
    333.91 dc20
                                                                                        96-35189
                                                                                                      CIP

ISBN   0 304 33567 3 (hardback)
            0 304 33568 1 (paperback)

Typeset by Ben Cracknell Studios
Printed and bound in Great Britain by
Biddles Ltd, Guildford and King's Lynn

Contents

Preface and Acknowledgements		vii
Chapter one	Sharing a Common Good	1
Chapter two	Avoiding the 'Tragedy of the Commons'	15
Chapter three	Hydraulic Societies and Regional Hopes	31
Chapter four	The Lure of the Dam	47
Chapter five	Fighting over Water	61
Chapter six	Small and Local	73
Chapter seven	The Women at the Well	83
Chapter eight	Marketizing Water	91
Chapter nine	The Unequal World of Water	107
Chapter ten	Dirty Water	115
Chapter eleven	A Confluence of Crises	123
Chapter twelve	The Delights of Water	133
Appendix	Information and Action	141
Index		143

Preface and Acknowledgements

> *There is a bond that links all men and women in the world so closely and intimately that every difference of colour, religious belief and cultural heritage is insignificant beside it . . . composed of 55 per cent water, the life stream of blood that runs in the veins of every member of the human race proves that the family of man is a reality. Thousands of years ago man discovered that this fluid was vital to him and precious beyond price.*
>
> **Richard M. Titmuss, *The Gift Relationship*[1]**

Richard Titmuss was not writing about water but about blood. He was making a study of blood transfusion and its implications, comparing the commercial market in bought blood with the voluntary donation of blood. He found that the dominant characteristic of the American blood-banking system was a redistribution of blood and blood products from the poor to the rich, since, having nothing else to sell, the people who sold their blood tended to be the unskilled, the unemployed and other 'low income groups and categories of exploited people'.[2] He found that when voluntary blood donors in Britain were asked about their motives, 'the vividness, individuality and diversity of their responses add life and a sense of community to the statistical generalities' but that 80 per cent of answers 'suggested feelings of social responsibility towards other members of society'.[3]

He concluded that on four testable non-ethical criteria, the commercial market in blood was bad:

> *In terms of economic efficiency it is highly wasteful of blood; shortages, chronic and acute, characterize the demand and supply*

> *position and make illusory the concept of equilibrium. It is administratively inefficient and results in more bureaucratization and much greater administrative, accounting and computer overheads. In terms of price per unit of blood to the patient (or consumer) it is a system which is five to fifteen times more costly than the voluntary system in Britain. And finally, in terms of quality, commercial markets are much more likely to distribute contaminated blood . . .*[4]

Titmuss died in 1974, and consequently did not live to see the transfer of the market ideology from economic theory to political dogma in Britain. Nor did he live to see the disaster that befell haemophiliac patients, heavily dependent on blood products, as a result of the importation of contaminated blood. His motive was simply to dispute 'the philistine resurrection of economic man in social policy'.[5]

Blood, as the saying goes, is thicker than water. Blood is an individual possession, water is a common necessity. But there are parallels between the two, since it is water that holds together the constituents of blood and is equally vital for survival. As Michael Allaby explains,

> *An adult human must consume at least 1.75 pints (one litre) of water a day, either by drinking or as an ingredient of food, and a person deprived of water will survive for no more than about six days, or in a hot environment perhaps for as little as two or three. A healthy person can survive for several weeks without solid food.*[6]

Water is as vital to us as blood, and is thus, in the words that Titmuss used, precious beyond price. Knowledge of this explains the outrage we feel on learning that households in England have been deprived of a water supply through non-payment of charges imposed by private companies to which the government has sold a community resource. And hence the anguish we should feel when we are told that a third of the world's population does not have access to a supply of safe drinking water, and that a third of all deaths in the world every day result from waterborne diseases.

Somebody has to restate the self-evident fact that water, a continually renewed but not inexhaustible resource, belongs to everybody, but not to any particular body of people who have chosen to control its supply. This is not to say that those who deliver it should not be rewarded. The water-seller belongs to an ancient service industry. Indeed, there are landless peasants in Bangladesh who have found a livelihood in a co-operative venture to supply water through mobile equipment to farmers whose land is too dispersed for mechanical irrigation technology.[7]

But water is also essential for the production of everything we eat and drink and use, for every form of industrial production, and for every kind of cleanliness and comfort. Like blood, it is too precious to be seen as a commodity. It is also the material of endless human delight, as we know from our pleasure in rivers and streams, ponds, lakes and the seaside and the fountains and swimming pools to be found in every town and city.

This book is not an attempt to repeat for water the study that Titmuss made of blood as a transfer of assets from the poor to the rich or as a gift from the fortunate to the unfortunate. It simply seeks to give a short and simple account of the immense social issues raised locally and globally by our universal need for water, and by the various water crises now facing the world.

I am deeply indebted to the great variety of authors and researchers whose findings and opinions I have quoted. Full details of their publications are to be found in the notes at the end of each chapter of the book. I must also thank William Heinemann Publishers for permission to quote from Alison Lurie's *The Nowhere City*; Weidenfeld and Nicolson for permission to quote from Joan Didion's *The White Album*; and Oxfam Publications, Earthscan Publications and Intermediate Technology Publications for similar kindnesses. I want to draw the reader's attention to the existence of various organizations with a continuous worldwide concern over water issues, listed in the Appendix.

I am also grateful to a variety of writers whom I have never met, who have worked away for years at our dilemmas over water. I am thinking especially of Fred Pearce in London and Jean Robert in Mexico, and of Bharat Dogra of the South Asian People's Environmental Network in New Delhi, where he edits *News from Fields and Slums in India*.

If you probe deep enough, everyone in the world has opinions about water and its fair distribution, but I have a particular debt to various specialists who have pointed me in the direction of facts I was seeking, and especially to Edmund Penning-Rowsell of Middlesex University and Paul Herrington of Leicester University. I am most grateful for the kindness of Anne McLean in Toronto and Neil Birrell in Dorset for translating source material for me.

Needless to say, none of these people are responsible for the uses I have made of their wisdom, but all are thanked.

COLIN WARD
April 1996

Notes

1. Richard M. Titmuss, *The Gift Relationship: From Human Blood to Social Policy* (London: Allen & Unwin, 1970), p. 15.

2. *Ibid.*, p. 246.

3. *Ibid.*, p. 235.

4. *Ibid.*, p. 246.

5. *Ibid.*, p. 14.

6. Michael Allaby, *Water: Its Global Nature* (London & New York: Facts on File, 1992), p. 7.

7. Geoffrey D. Wood and Richard Palmer-Jones, *The Water Sellers: A Co-operative Venture by the Rural Poor in Bangladesh* (London: Intermediate Technology Publications, 1991).

CHAPTER 1

Sharing a Common Good

Human settlements began where people saw the chance of establishing sustainable lifestyles on fertile land where water was more or less regularly available. Thus the inhabitants came to create social arrangements for the sharing of land and water for various purposes within their own communities.

Because it is fixed and stable, land can be divided by hedges and walls. Thus land has the potential to be held in common or to become, as it has done for many countries, the foundation of private property, personal wealth and inheritance. By contrast, water has to be a communal asset because it will not stay still. For thousands of years, legal systems across the world have accepted and insisted that there can be no ownership of running water.

David Kinnersley, *Troubled Water* [1]

The British Isles are blessed with an equable climate that most people in the world would envy. I live on what is seen as the 'dry' side, in rural East Anglia, which has less than half the national average rainfall. Our house, like those of the neighbours, was connected to a piped water supply as recently as 1952. Until that year the most important items of water engineering for the occupants were the well and a collection of galvanized iron buckets, filled every day from the well for drinking and cooking, from tanks draining the roof gutters for washing, and from the pond fed by field drains for watering the garden and the animals. Water was never wasted, because every member of every family, apart from those who employed servants, knew the incessant labour of carrying it.

Ralph Whitlock, recalling life in the 1930s in rural Somerset, stresses the sheer physical effort involved:

> One of the most important items in Saturday's preparation for the Sabbath was the drawing of water by windlass from the village wells. Conventionally that was the responsibility of the man of the household on Saturday afternoons or evenings. Pulling up a ten-gallon bucket of water from a depth of eighty feet occupied, say, ten minutes, so most householders had to reckon on an hour or two's work at the end of the week, except, of course, when providential rain relieved them of the chore. Rainwater was perfectly satisfactory for washing, though hardly for drinking, unless boiled. Drinking water was stored in large earthenware pans, generally with a lid . . . In times of drought, drawing water was a major preoccupation for farmers, who had to employ a man virtually full-time to supply the needs of the farm livestock.[2]

The deeper the well, the purer the water, but the greater the labour. It is not surprising that, as the technology improved, the village pump became a focus of rural life. We use the expression 'parish pump politics' as a derisory way of describing small local issues, but the pump is a powerful symbol of community effort.

Indeed, Brian Bailey, the historian not of parish pumps but of village greens, suggests that

> One possibility which has not been proposed, as far as I know, is that small greens were conceived as central areas reserved for the protection of the common water supply and to give access to it. It is easy for us in these days of piped water to every home to overlook how the absolutely overriding consideration in the establishment of any village or hamlet was the availability of water. When a well was dug, with great labour, to supply all the villagers with their water, it must have been regarded with a protective reverence that we find hard to imagine today.[3]

Bailey describes a whole series of village pumps, like the one at Pound Green near Earls Colne in Essex, which has the inscription, 'This well was digged in thankful commemoration of the absence of cholera for the common use of the people to provoke them to cleanliness'; or the one at Stoke Row in the Chilterns, 368 feet deep and given to the village in 1864 by the Maharajah of Benares when told of the problems it had in getting a water supply; or the five village wells at Tissington in Derbyshire, decorated with flowers every year to celebrate the fact that they had 'protected the villagers from drought and disease by always supplying pure water'.[4]

He remarks how these communal pumps, though now obsolete are 'often still carefully protected from vandals and pollution. They stand like symbols of village continuity, and . . . the community's source of life.'[5] This is certainly true of my nearest village, where the pump survives, but is boarded up to protect it from those entrepreneurs who know that there is a market for those bits of Victorian cast-iron now known as 'country bygones'.

Apart from the endless labour involved, life before piped water was not an idyll of bucolic health. In the late 1940s George Ewart Evans, later famous as a rural historian, moved to Blaxhall in Suffolk, where his wife became the village schoolteacher. The whole family had frequent stomach troubles:

> *Most newcomers suffered in the same way. A cynic was of the opinion that you would suffer until you became 'manured' to it . . . Eventually we had the water in the school well analysed. This showed that the bacteriological findings were satisfactory; but the chemical content of the water was not investigated. Reports in the press had told us that very young babies developed cyanosis through being given well water in their food, and some had died. The offending chemicals were nitrates which were poured in tons on the soil of this intensely arable area as farm fertilizers.*[6]

Evans, who became clerk to the parish council, had to make himself 'an aggressive nuisance' to get piped water to Blaxhall, since the membership of district and county councils were wealthy people, isolated from the disadvantages that rural life involved. 'A farmer, for instance, was able to bore an artesian well from which he pumped water by means of a petrol engine; and by the same means he could generate his own electricity.'[7] Finally his village, at about the same time as ours, got a piped water supply.

In the wake of water on tap came the water-closet, discharging, as in our house, into a cesspool, the emptying of which used to be undertaken by the local authority, but which I now pay for individually to a one-man business. The contents of his tanker are eventually spread on land, not discharged into rivers and seas to pollute the beaches of the Eastern coast.

Naturally, a clean piped water supply and a sewage system came a lot earlier in the big towns, because of the ever-present fear of epidemic diseases. Our nearest big town, Ipswich, was well-endowed with springs which flowed down into the river Orwell. Its historian explains that the town's poorer people had to fill their pails from open streams or backyard wells:

> It is not at all surprising that a report in 1848 spoke of 'the foundations of the town saturated with foul and pestilential moisture' or that about the middle of the 19th century Ipswich had a death-rate as high as any English large town and higher than most. A waterworks was constructed by the Ipswich Water Works Company, but comparatively few houses were connected to the public supply until the Corporation bought the undertaking in 1892.[8]

Our local water history mirrors that of the country as a whole. A network of suppliers was built up, some run by local councils, some of them private companies but with a statutory limit on dividends. Slowly a water ethic was built up (very slowly in rural Suffolk) that saw water as a necessary common good rather than as a commodity. The recognition that easy access to a clean water supply was a basic human need was enshrined from the last century onwards by a requirement in the Public Health Acts which declared that a house without an adequate water supply was 'unfit for human habitation'.

Whether the distribution agency was municipally owned or a private company, the capital cost of underground pipes was met as a public duty, and the cottager was charged with a few pence on the rent or the rates (a local tax based on the size of a house). For poor families every recurring expense was a problem, but the water rate was a trivial charge, since water was everyone's natural right. Farmers, like industrial users, paid a commercial rate for a metered supply.

There is a direct relationship between the size of a human settlement and the need for safe water and safe sewage disposal. The huge expansion of the urban population was accompanied by dysentery, typhoid fever and cholera, and by the need for large-scale water engineering and for 'sewage farms' discharging a safe effluent from urban drainage systems. The spirit of the time believed that private enterprise would solve all human problems, while public enterprise was wasteful and wicked. The result was the new crime of water-stealing. Nigel Morgan, the historian of one of the new industrial towns, Preston, in Lancashire, has excavated from its local paper for the year 1844 the evidence of this crime. One report, headed 'Caution to Water Stealers', says that

> *Yesterday, Elizabeth Stubbs appeared before the magistrates at the Town-hall, charged with taking water from one of the taps supplied by the Preston Waterworks Company, she not having made any contract with the said company to do so. The case being fully proved against her, she was ordered to pay a small penalty and costs.*

Another letter to the editor, signed by 'One Who Wants the Water Tax Taken Off', explains that

> *I had occasion to go down a court where there are a number of cottages, but all without any supply of water. 'How do you get water?' said I to a widow, being in one of the houses. 'We steal it,' was her reply. 'We get it from _____ , and we run as far as we can for fear of being seen.' Another argument, thought I, in favour of cheap water. Cottages paying 1s. 3d. per week rent would not mind paying a 1d. per week for water; but can such be expected to afford 12s. a year?*[9]

This unknown correspondent of a century-and-a-half ago had grasped a truth that the British have to face today. It is overwhelmingly in the interest of the whole community that everyone should have access to clean water, regardless of their ability to pay for it. The experience of Preston, with its appalling mortality figures, was that the solution, as in Ipswich, was for the local council to purchase the Waterworks Company which supplied a minority of households. In Liverpool, in the year that saw the prosecution of water-stealers in Preston, Samuel Holme was arguing that

> *water is as essential to the health and comfort of mankind as the air we breathe, and when mankind congregate in masses counted only by*

> tens of thousands, it is essential to the public health that it should be most abundant, not doled out to yield 30 per cent interest, but supplied from the public rates and at net cost.[10]

Most secondary-school students of history learn of the struggle for public health, clean water and efficient sewage disposal in nineteenth-century London, and of the remarkable detective work of Dr William Budd and Dr John Snow in identifying the origins of the cholera epidemics that plagued London and other cities. Anthony Wohl summarizes the exciting achievement of Snow's careful observations thus:

> In 1854 Snow was given the opportunity to prove his theories when he dramatically and conclusively traced cholera deaths to houses supplied by the suspect water of the Southwark and Vauxhall water company. When he managed to persuade the local authorities to lock the handle of a pump in Broadwick Street in Soho (a compact area where over fifty people a day were dying of cholera) the deaths there came to a sudden halt, and although it was not until 1883 that Koch succeeded in isolating the cholera bacillus, Snow's work marked a triumph for the young science of epidemiology.[11]

It is difficult today to appreciate the importance of Snow's findings. For at the time of the terrible cholera outbreaks in that mid-century it was the poor who were blamed, for their well-known ignorance, lack of hygiene and general moral depravity. Just before the second great outbreak of 1848–9, the Metropolitan Sanitary Commission warned that 'where people live irregularly, or on unsuitable diet and at the same time filthily', they must expect to perish. Bill Luckin, the historian of the River Thames, who cites this finding, remarks on the change in the attitude of scientists and medical investigators by the end of the nineteenth century; 'To them, at least, whatever may have been their doubts about the explicitly political implications of municipalism, interventionism and collectivism, "the salvation of the city" was nothing less than a binding moral duty.'[12]

This binding moral duty to ensure that every household had access to a clean water supply and safe disposal of wastes, was slowly accepted everywhere. It came rapidly to the cities, where the consequences of failing to acknowledge this duty were obvious.

Hence the great monuments of Victorian water and sanitary engineering that astonish us today. It came, as we have seen, much more slowly to remote rural areas where the risks and the voters were fewer. Twentieth-century population growth, the spread of suburbs and our changing habits brought an increasing demand for water, and the dominant ideology of the suppliers was not the financial priority of reducing costs, nor the ecological priority of conserving the more vulnerable sources, but the engineering priority of, as a matter of pride, meeting demand at all costs.

The world of sewage disposal was similarly dominated by engineering prowess. The manager of the sewage works was proud to be photographed drinking a glass of the crystal-clear final effluent, or would happily draw attention to the fact that water consumed by Londoners had, before being extracted from the Thames, passed through households in places like Maidenhead, Reading and Oxford. This was an immense long-term achievement when we consider the pestilential state of British rivers in the last century.

But what of the solid wastes left behind? Here, as with the growing problem of the sheer volume of other forms of domestic waste, there were, and are, several points of view, all of which mingled technical, ideological and economic considerations. In the late 1960s the British government commissioned working parties to report on both.[13]

It was found that of the total sludge produced at inland sewage works in England and Wales, two-fifths was applied to agricultural land, two-fifths was disposed of in other ways, including composting with domestic refuse, and roughly one-fifth was dumped at sea. The working party on sewage noted that

> For many years estuaries have been regarded as capable of taking virtually unlimited polluting loads – probably because it was erroneously believed that their salt content had special purifying powers. Consequently effluents have been discharged without treatment, which has resulted in very severe pollution especially where large conurbations exist on, or near, the mouths of rivers.[14]

The 1970 working party similarly noted with horror that the sewage and trade waste of a population of about 6 million was discharged directly into the sea or to estuaries with only partial or no treatment. It listed the effects on health and amenity and on marine life, stressing that 'We cannot accept that conditions on our beaches should be below the standards of hygiene and decency that we expect in our homes, streets and workplaces.'[15]

This was one of the many issues relating to water which induced the British government to introduce its Water Act in 1973, bringing what was thought to be a coherent policy to water supply, river management and sewage disposal. As Fred Pearce put it in his study of Britain's water crisis:

> *In 1974 the new broom of local government reorganization swept away 100 water boards, 50 local council water undertakings, seven water committees, 27 river authorities, two river conservancies, 1,366 council sewerage undertakings and 27 joint sewerage authorities and replaced them with 10 regional water authorities to cover the whole of England and Wales. The new authorities had power over the whole 'water cycle' from upland reservoirs to seagoing sludge ships; from land drainage to water mains and from pollution control to flood prevention. The only survivors from the old system were 30 water companies, which were saved by Conservative ministers. The new authorities are strange hybrid bodies, neither nationalized industry, nor local government . . . And behind them all is a National Water Council appointed by Ministers . . .*[16]

For those people who believed that the problems of the water industry and of pollution were the result of fragmentation of control and of the separation of river management from water extraction and sewage management, the new unified system, and the prospect of vast new investment, was an ideal solution. Even the remaining private water companies became statutory agents of the new water authorities. But the results were disappointing. Naturally the first act of any huge new public body is to build itself a new headquarters on a scale commensurate with its status and its responsibilities. But its second act is to take over all the expansive future plans of its constituent bodies.

Forecasters and planners in any field predict that current trends will go on for ever. Fred Pearce provides an illuminating account of the habit of thinking big in the water industry:

> *In 1971 some 14,100 million litres of water left treatment works in England and Wales. That was 35 per cent more than in 1961 and 135 per cent more than in 1940. There was almost universal agreement that the trend would continue. Few planners thought to examine what their water was used for, who was using it and whether they were likely to carry on wanting more . . . The Water Resources Board, shortly before it was wound up in 1974, predicted that demand for water would double by the year 2000 . . .*[17]

Then came the Middle East oil crisis, which first doubled and then trebled the price of oil and was followed by recession, 'coupled with a "save it" outlook which affected everything . . . Industry's demand for water has fallen every year since.'[18] And Pearce goes on to explain that

> *Most of the 'new men' occupying larger desks and drawing larger salaries inside the new regional authorities were the bigger fish from the old council sewerage boards and river authorities. They had brought their big project ideas with them. But, almost as soon as they had collected their keys for the executive washrooms, they found that long-cherished projects might fall victim to lower demand forecasts. Many simply refused to believe the predictions. Some saw them as an extension of the conservationists' campaign against the new reservoirs. In the confusion a lot of things got built which should have been prevented.*[19]

The new authorities neglected the laborious and unexciting task of repairing the crumbling Victorian structure of pipes, sewers, pumping stations and processing plants, quite apart from the coastal horrors of sea pollution. No one blames the engineers of years gone by for failing to anticipate that 40-ton trucks would be allowed to pulverize vital services beneath the roads without recompense, but everyone knows that iron corrodes. And then the money ran out. In 1982 the government was permitting the water industry to spend only half the sum it had put into capital investment in 1974.

When the Conservative government came to power in 1979 no one anticipated that one of its achievements would be to change the nature of water from a common good to a commodity. However, ten years later, water joined the other publicly owned utilities in being offered for sale to a public which already owned it collectively. The historians of another industry remind us that the privatization of water was not so easy, since the ministers responsible were 'defeated in the House of Lords and threatened with prosecution by Brussels on water quality standards; attacked by environmental groups on the same issue and on the fate of the water authorities' massive land holdings.'[20] Nevertheless,

> *The flotation, in November 1989, was successful to the extent that the shares were 5.7 times oversubscribed. However, the amount of the net proceeds to the Treasury of the sale was negative: £5.3 billion was raised by the share issue, at a cost of £5 billion debt write off, £1.6 billion cash injection to the authorities and £100 million flotation expenses, leaving the Government £1.4 billion out of pocket on the deal.*[21]

The water-using public – which means every household in the country – was probably unaware that a priceless asset it owned had been sold at a loss in pursuit of a transient governmental ideology, but it rapidly became aware of the consequences. Every householder witnessed an increase in water bills by an average of 67 per cent between 1989–90 and 1994–95, while the companies' profits rose on average by 20 per cent a year from 1989–90 to 1992–93, and profit margins rose from 28.7 per cent to 35.6 per cent.[22] Disquiet over water prices has grown with continual reports of the vast sums and share options that the directors and executives of the water companies have paid to themselves, by revelations that their diversifications outside the water industry have been a financial disaster, and that far less has been spent on the upgrading of plant and installations than the public was led to believe.

We can take all this for granted as an aspect of the enterprise economy. My concern is with the situation of the poor. The National Consumer Council's report on water pricing found that in the highest-charging area of Britain, that of South West Water, 'the average water bill takes 4.9 per cent of income for a household of two adults and two children, 7.6 per cent for a lone parent and child, and 9.1 per cent for a single pensioner' and it comments that 'These figures represent a substantial financial strain on households that are least able to afford price increases for essential services.'[23]

Where I live, tenants of local authorities paid a small sum in their weekly rent for water, once controlled by the same council. Today the council declines to be a collecting agency for a private company and the vastly increased water bill, payable in advance, has become another of the overhead costs of living which poor people have, somehow, to budget for. Until 1988, people in receipt of the government's 'supplementary benefit' had their water bills paid, but with the abolition of that payment they too became responsible for finding the money. Privatization of water supply brought a new aggressively commercial approach to the poorest of water-users. Thousands of households had their water supply cut off, an action that I had imagined to be illegal, as it is in Scotland and Northern Ireland. I was wrong, and the representative of Thames Water told the press in 1992 that 'We are being too soft, and that is why our disconnections level will rise.'[24] A handful of members of parliament, led by Helen Jackson, have sought to introduce legislation to require the water companies to recover outstanding payments through the courts, like any other creditor, and not by disconnection.[25] They have not succeeded, despite their accumulation of countless case-histories of disadvantaged people penalized by the ruthless policies of water companies.

At a meeting Helen Jackson convened in 1993, John Middleton, director of public health for the Sandwell Health Authority in the West Midlands, drew attention to the collapse of a sense of public morality since the sanitary campaigns of 150 years ago, recalling that 'the Victorians at least recognised the need to provide safe, wholesome water supplies for everybody, rich and poor. Water disconnection is something we should not tolerate in a civilised society.'[26] He said that during the period of 1991 and 1992, with a marked rise in disconnections in his area in which over 1,400 households lost their water supply, 'over this period cases of hepatitis and dysentery rose tenfold'.[27] When the British Medical Association examined the same issue it found that there were a number of vulnerable groups for whom a guaranteed water supply was vital[28]. They were people with medical conditions requiring the use of additional water for bathing or washing clothing, young children and elderly people. When the Policy Studies Institute initiated a careful study of water debt and disconnection it found that 'Of these, only elderly customers seemed to have a low risk of water debt and hence of disconnection.'[29] It also reported that 'During 1994 about 2 million households fell into arrears with their water bills and 12,500 ended up disconnected from the water supply.'[30]

Now personally, quite apart from medical arguments, I cannot imagine a situation where I could survive without a water supply, and nor can any reader. We all need to drink and prepare food, we all produce faeces and urine and need to dispose of them, we all need to wash. To deny any of us access to water puts us in the situation of the woman whose prosecution in Preston I have mentioned, who was found guilty of stealing water.

That was 150 years ago, and it is sobering to realize that, in civilized Britain, the fantasies of wealthy advocates of the logic of the market should reduce us to the brutality of that kind of attitude. The modern equivalent of Elizabeth Stubbs is in one sense worse off than she was. For a century ago, as we have seen, every village had its parish pump, provided either by community effort or philanthropic concern, accessible to all. A drinking fountain and basin in East Street, Colchester, now dry, bears the inscription, 'With Joy Shall Ye Draw Water', and all elderly Londoners will remember the innumerable watering points provided for humans and animals alike by the Metropolitan Drinking Fountain and Cattle Trough Association. Access to water was recognized as a universal human right rather than a commodity.

At Blockley in Gloucestershire, an archway in a wall protects a spout from a natural spring, and incised in the pediment above is the legend, 'Water From the Living Rock: God's Precious Gift to Man'. It is called the Russell Spring after the family that connected it for public use 150 years ago, 'at a time when cholera was rife in the area'. In 1994 a newly-formed Cotswold Water Company applied for an abstraction licence from the National Rivers Authority to bottle and sell 750 gallons a day.[31] The anecdote provides a parable of the coarsening of British sensitivity to the nature of a universal human need.

Notes

1. David Kinnersley, *Troubled Water* (London: Hilary Shipman, 1988) p. 1.

2. Ralph Whitlock, *The Lost Village: Rural Life Between the Wars* (London: Robert Hale, 1988), p. 170.

3. Brian Bailey, *The English Village Green* (London: Robert Hale, 1985), p. 22–3.

4. *Ibid.*, p. 86.

5. *Ibid.*, p. 106.

6. George Ewart Evans, *The Strength of the Hills* (London: Faber & Faber, 1983), p. 159.

7. *Ibid.*, p. 160.

8. Robert Malster, *Ipswich: Town on the Orwell* (Lavenham: Terence Dalton, 1978), p. 71.

9. Nigel Morgan, *Deadly Dwellings: The Shocking History of Housing and Public Health in a Lancashire Cotton Town* (Preston: Mullion Books, 1993), p. 14.

10. Derek Fraser, *Power and Authority in the Victorian City* (Oxford: Basil Blackwell, 1979), p. 145.

11. Anthony S. Wohl, *Endangered Lives: Public Health in Victorian Britain* (London: J. M. Dent, 1983), p. 191.

12. Bill Luckin, *Pollution and Control: A Social History of the Thames in the Nineteenth Century* (London: Adam Hilger, 1986), p. 182.

13. *Taken for Granted*, Report of the Working Party on Sewage Disposal (London: HMSO, 1970); *Refuse Disposal*, Report of the Working Party on Refuse Disposal (London: HMSO, 1971).

14. *Taken for Granted*, para 182.

15. *Ibid.*, para 228.

16. Fred Pearce, *Watershed: The Water Crisis in Britain* (London: Junction Books, 1982, pp. 3-4.

17. *Ibid.*, p.15.

18. *Ibid.*, p.17.

19. *Ibid.*, p.17.

20. Jane Roberts, David Elliott and Trevor Houghton, *Privatising Electricity: The Politics of Power* (London and New York: Belhaven Press, 1994), p. 35. See also John Ernst, *Whose Utility?: The Social Impact of Public Utility Privatization and Regulation In Britain* (Buckingham: Open University Press, 1994).

21. Roberts *et al.*, *op. cit*, p. 35.

22. National Consumer Council, *Water Price Controls Review: Key Consumer Concerns* (London: NCC, 1994), p. 20.

23. *Ibid.*, p. 7.

24. *Guardian*, 2 September 1992.

25. *Hansard* (House of Commons Official Report), 25 November 1992, 28 January 1993, 25 February 1994.

26. *Independent*, 29 January 1993.

27. *Ibid.*

28. British Medical Association, *Water: A Vital Resource* (London: BMA, 1994), p. 16.

29. Alicia Herbert and Elaine Kempson, *Water Debt and Disconnection* (London: Policy Studies Institute, 1995), p. 6.

30. *Ibid*, p. 8.

31. Paul Stokes, 'Villagers battle to save spring from the bottle', *Daily Telegraph*, 19 October 1994.

CHAPTER 2

Avoiding the 'Tragedy of the Commons'

The water communities built up on the bases of the rivers Genil, Segura, Júcar, Turia, Mijares, Jalón, Ebro and several other rivers of the Peninsula, are admirable examples of solidarity and social co-operation of a truly marvellous delicacy and perfection, unequalled by the most complex works of precision engineering. In their superb rhythms human handiwork rivals that of Nature. They were the unique creation of the Spanish people on foundations laid by the Phoenicians, the Roman Empire, and finally the Muslims.

Joaquin Costa, *Agrarian Collectivism in Spain* [1]

History is full of examples of societies where the rich were clean and the poor were dirty, and where the rich ate well-watered fruits while the poor ate rye bread, simply because the time and energy of the poor were spent in attending to the needs of the rich. But there is also a long history of human societies that developed elaborate systems of ensuring fair access for all to the means of life and livelihood, including water. The modern technology of pipes, pumps and motive power, quite apart from eliminating back-breaking labour, should make fair shares of common necessities easier to ensure, but instead common ownership is seen as an encouragement to greed and wastefulness. How did people reach such a misanthropic conclusion?

In 1968, Garrett Hardin, professor of biology at the University of California, published an article which attempted to come to terms with the human dilemma which he expressed in a parable as 'the tragedy of the commons'.² Imagine, he suggested, an ancient common pasture on which every herdsman grazed his animals. It could work reasonably well for centuries, he argued, 'because tribal wars, poaching and disease keep the numbers of both man and beast well below the carrying capacity of the land. Finally however, comes the day when the long-desired goal of social stability becomes a reality. At this point, the inherent logic of the commons remorselessly generates tragedy.'

The reason is, he suggested, that each herdsman will pursue his own interests by increasing his herd, while the common land will become devastatingly overgrazed:

> Each man is locked into a system that compels him to increase his herd without limit – in a world that is limited. Ruin is the destination toward which all men rush, each pursuing his own best interests in a society that believes in the freedom of the commons. Freedom brings ruin to all.

Hardin's neo-Malthusian approach, declaring also that 'freedom to breed will bring ruin to all', was widely criticized from several standpoints. For the authors of *The Little Green Book*, the implications of his 'living in a lifeboat' theory were shocking: 'Rather than being on a lifeboat, are we not on a larger liner, in the luxury of the Captain's stateroom, while the masses are starving in steerage?'³ To this important point, Hardin made a pre-emptive response, since he noted that

> We must admit that our legal system of private property plus inheritance is unjust – but we put up with it because we are not convinced, at the moment, that anyone has invented a better system. The alternative of the commons is too horrifying to contemplate. Injustice is preferable to total ruin.⁴

The American advocate of 'bio-regionalism', Kirkpatrick Sale, found that Hardin's parable suggested an important and unquestionably tragic truth:

> The oceans continue to be overfished, particularly by Russian and Japanese fleets, because it is in the immediate self-interest of each nation to take as much as it can from the sea – particularly if there is a threat of eventual depletion – even though every nation knows that

at some near time there will be no more fish at all to catch. American steel manufacturers continue to resist measures to limit the poisonous emissions they pour into the air each day, because it is in the interest of each company to avoid the expense of pollution controls and redesigned furnaces for as long as possible, although steel company executives know that they and their families and workers and townsfolk have to breathe the resultant polluted air and that this is very likely to cause severe illness and early death. The American West is becoming desertified, losing perhaps 10 million acres of grassland a year, for the simple reason that the ranchers of the area, acting out Hardin's scenario, continue to overgraze their herds in this sparse territory despite a decade-long policy of the Bureau of Land Management to curtail them.[5]

Kirkpatrick Sale blames not population pressure, but several related factors. He sees the situation as an aspect of competitive capitalism which ensures that 'those who are operating out of *self-interest*, do not see and cannot feel a *communal-interest*'.[6] This results in an inability to see the need for a steady-state economy and the centrality of the concept of ecological balance:

> *The shepherd would know the limits of the commons and its importance to his family, to his children and their children, to his neighbours, to the past and future of his community, and that over-riding communal-interest would easily outweigh the possible personal gain of putting another animal out to feed.*[7]

But he recognizes that this too, is a function of the *scale* at which decisions are made:

> *There can be no communal-interest among 200 million people, or 20 million, or even 2 million, because there is no way for the human heart with all its limitations to perceive the interconnectedness of all those lives and their relevance to its single life; we cheat on our income tax and drive at 65 mph, and ignore beggars on the street because we perceive no community at the scale at which we live. Nor can there be communal-interest over distances of 3,000 square miles, or 300 square miles, or even 30 square miles, because there is no way for the human mind in all its frailty to conceive the complexity of an ecosystem so large and its single place within it . . . Only when the shepherd knows*

his world and the people in it and feels their importance to his own well-being, only when he realizes that his self-interest is indeed the communal-interest, will he voluntarily limit his flock.[8]

The fable of the tragedy of the commons has been used as the final word in a dozen different environmental arguments. There is every reason to take warnings of over-population seriously, but they come with a strong whiff of hypocrisy from public people with a long life-expectancy in the rich world, when addressed to poor people with a pathetically short life-expectancy in the poor world. As Peter Marshall puts it,

Industrialised countries should not call for population control elsewhere but should first set their own house in order and consume less. In particular, they should not suck up the world's non-renewable resources in the form of primary materials and then through advertising create a market for their manufactured goods . . . In the long run, the most important means to check excessive population is the prospect of a decent life. If life is secure and pleasant, families become smaller.[9]

Marshall is eloquent on the real issues involved:

Like other animals, humans are essentially dependent on plants as a source of food, and ultimately numbers will be limited by the carrying capacity of the world defined in terms of the availability of food. But there are enormous individual and regional differences. Two thirds of the world's human population live in poverty while the remaining third enjoy comparative luxury. The poor consume a small amount of the world's non-renewable resources but their growth in population is double that of those living in developed countries. The average European consumes 600 times more steel than the average African; if everyone used oil like the average American, there would be no oil left on the planet in a few years . . .

As Europeans colonised the world, they brought with them their culture as well as their technology, a culture which saw 'development' and 'progress' in terms of technological mastery over nature. The message was simple and stark: the world is there to be exploited and despoiled. But the Western way of life can never be adopted universally because there are simply not enough resources to go round.

> *The poor countries of the world have developed a taste for consumer goods which they will never be able to satisfy – except for their elites. So the cycle of famine and misery will continue unless there is a fair distribution of resources, less consumption in the richer countries, and a move towards a sustainable global economy which recycles materials and uses renewable energy supplies. If not, then those in the West will probably continue to sit in front of colour television sets, eating processed meals in centrally heated houses, and watch images of others starving to death transmitted 'live' via satellites orbiting the earth.*[10]

Garrett Hardin also claimed that the tragedy of the commons reappears in a reverse way in the problems of pollution, seen once again as a population question. His grandfather used to tell him that flowing water purifies itself every ten miles; and

> *the myth was near enough to the truth when he was a boy, for there were not too many people. But as population became denser, the natural chemical and biological recycling processes became overloaded, calling for a redefinition of property rights.*[11]

Since he concluded that it was a rational individual decision by each herdsman to enlarge his herd, he was bound to make the same calculation on pollution:

> *The rational man finds that his share of the cost of the wastes he discharges into the commons is less than the cost of purifying his waste before releasing them. Since this is true for everyone, we are locked into a system of 'fouling our own nest', so long as we behave only as independent, rational, free-enterprisers . . . Indeed, our particular concept of private property, which deters us from exhausting the positive resources of the earth, favours pollution. The owner of a factory on the bank of a stream – whose property extends to the middle of the stream – often has difficulty seeing why it is not his natural right to muddy the waters flowing past his door. The law, always behind the times, requires elaborate stitching and fitting to adapt it to this newly perceived aspect of the commons.*[12]

However, the aspect of his famous essay that stays in people's minds is his conclusion that 'the tragedy of the commons as a food basket is averted by private property, or something formally like it'.[13] It is always raised as the final argument against community ownership and control of resources. In Britain it is even used as a belated justification for the process of Enclosure of the common fields, common lands and wastes, which was a continuous process over centuries, culminating in the Parliamentary Enclosures of, say, 1750 to 1850. Different schools of historians have been arguing for a century about the Enclosures and their effects. One of the most recent scholars, Dr K. D. M. Snell, considers that current research and reappraisal of open-field agriculture has established that 'the open fields were far more open to innovative and flexible agriculture than once supposed', and that 'the account of them as seriously backward and by nature inhibitive of new technique is most certainly incorrect'.[14]

A team of investigators were appointed by the United Nations Environment Programme to report on the alleged downward spiral of the Himalaya region's forests, where it was feared that the rate of timber use had overtaken the rate of new growth. They described what they saw as Garrett Hardin's law of the tragedy of the commons working in reverse in some Himalayan countries. They learned that for centuries the Sherpas

> *managed their common forest resources with the help of their social institution of forest guardians — a rotating office within each village, the annual holder of which, after due (but fairly casual) consultation, laid down the permissible extraction rates for fuelwood and constructional timber and enforced traditional fines on those villagers who did not comply.*[15]

But in the 1950s the forests of Nepal were nationalized and controlled by regionally based officials. The investigators found that the old system worked quite well, but that the new centralized system does not. Local, popular, control is the surest way of avoiding the tragedy of the commons.

Hardin, in fact, fully admitted this in the case of water, for he remarked, in a passage that those who use his argument to support private monopolies seldom quote, that

> *To many, the word coercion implies arbitrary decisions of distant and irresponsible bureaucrats; but this is not a necessary part of its meaning. The only kind of coercion I recommend is mutual coercion, mutually agreed upon by the majority of the people affected.*[16]

And in those circumstances where access to water is not determined by private or public exploiters of the universal need, this turns out to be true. Local control of water in those parts of the globe where its prime importance is for irrigation, is effected in an endless variety of popular and mutual systems. Robin Clarke describes many instances:

> *The classic example must be Bali, where the most sophisticated form of village level irrigation has existed for centuries. All farmers who take water from the same stream or river are members of an organisation called a* sebak, *which meets every 35 days and has its own system of law. This organisation plans planting times, distributes water equitably and fines those who cheat. The distribution system is that each 0.35 hectares of land is entitled to a* tektek *of water – the amount of water than can be delivered by a gap four fingers wide which is cut into the trunks of the coconut trees used to deliver the water.*[17]

The country whose history illuminates a whole series of aspects of the debate on 'the tragedy of the commons' is Spain. It had systems of common distribution of water resembling Robin Clarke's account of Bali, and it had examples of communality of land, of every kind of private ownership from *latifundias* to *minifundias*, and a landless peasantry living in hopeless poverty. The historian Raymond Carr described the vast private estates:

> *Whether the estate was farmed by its owner or let out to a rich tenant, the land, apart from a handful of permanent workers, was worked by gangs of seasonal labourers – the* braceros *who, unemployed for half the year, starved in the agro-towns. They were the most wretched agricultural labourers of Western Europe . . . A vast estate, using techniques which had scarcely changed since Roman times, still yielded an adequate income to allow its absentee owner to live a life of conspicuous consumption in the provincial capital, if he were a local* senorito, *or in Biarritz or Madrid if he was a great aristocrat.*[18]

This was one extreme of the private property recommended by Garrett Hardin, but the huge diversity of the Iberian peninsula provided examples of the opposite extreme. The geographer Ruth Way explained that

> *In the fertile Valencian huertas, where the hoe and spade are as important as the plough, 5 to 10 acres is the normal family unit. In the north and west, where the ancient Celtic law of equal distribution of land to inheriting sons still applies, farms become minute (¼ to 2 acres) as they are progressively split up, so much so that many men emigrate in order to provide land for the rest of the family.*[19]

In a country where private land-ownership ranged from *latifundias* of over 12,500 acres (5083 ha) down to *minifundias* of less than an acre (0.4 ha), neither of which could support the poor peasants dependent on them, there has always been advocacy of a communality of land. It came from historians describing the past and from prophets of the future. One of the celebrated members of the nineteenth-century movement known as Regenerationism, famous for his slogan 'School and Larder', was Joaquín Costa (1846–1911).[20] He was a prolific self-taught propagandist from rural Aragon, whose programme based on land redistribution, education and irrigation, appealed both to reformers from the political right and to anarchist revolutionaries on the left, as did his most celebrated book on agrarian collectivism in Spain.[21]

Irrigation of crops, in a land subject to irregularities of rainfall and prolonged droughts was practised in Spain through the building of barrages and the digging of canals in the Bronze Age, and was extended by the Romans for whom Spain was indeed a larder. (The word *huerta* for the amazingly productive lowland farms comes from the Latin word for garden, *hortus*.)

The centuries of Moorish occupation of large parts of the country (711 to 1492) added to the development of effective irrigated cultivation and introduced new crops: oranges, pomegranates and rice. Reconquest was a matter of imperial grandeur, not of good husbandry. Ruth Way explains that

> *The new Christian conquerors were soldiers, not farmers, and considered manual labour undignified, a sentiment still to be found in Spain. Only in those regions of adequate rainfall (north and west) or where local irrigation schemes survived (some areas of the east and south) was agriculture practised on an economic basis. Many areas were simply abandoned, to suffer the consequence of soil erosion, or allowed to revert to a degenerate type of scrub.*[22]

She describes how the irreconcilable political differences in Spain and the lack of capital for large-scale water engineering worsened the decline of rural Spain. Deforestation, under private ownership, worsened the water situation. Far from conserving forests, in a rational approach to woodland management, the desire for immediate financial returns decimated them. As Raymond Carr reported,

> Between 1866 and 1932 perhaps half the woods left standing by mid-century speculators were cut down. Without adequate cover, the soil of Spain blew away or was washed away by torrential winter rains; denuded watersheds, unable to absorb water in the higher reaches, produced disastrous floods or dried up in river courses in summer. Useful rivers, for irrigation and, later, for hydro-electric schemes, were a corollary of useful forests.[23]

This was not a tragedy of the commons, it was a tragedy of unrestrained individual self-interest, and it was the fate of the programmes of visionaries like Costa that the only twentieth-century Spanish governments to embark on large-scale but 'uneconomic' policies of re-afforestation were those of the authoritarian political right: those of Antonio Maura in the early years, of Primo de Rivera in the 1920s, and of Francisco Franco after the civil war of 1936 to 1939.[24]

Like all the rulers of poor countries, they thought on a large scale but did not threaten big landowners. At the level of the small cultivator, they failed to notice, as Raymond Carr did, that

> the small farmer could not modernise because he lacked the capital when his liquid resources were taken up by taxation and when credit at reasonable rates was denied him; he stuck to the noria – the water wheel with its pots turned by a mule – because he could not afford a pump.[25]

At the level of the great landowners, so long as they could derive an income from their peasants, they disdained this vulgar concern with water. Their approach was tellingly described by Edward Malefakis, in describing southern Spain, which (apart from Granada, where Moorish irrigation systems survived), even though water management could increase crop yields by from three- to sixfold, was the least irrigated region of Spain:

> *Because they refused to cooperate with the Hydraulic Con-federations by which Primo de Rivera attempted to expand irrigation facilities in the 1920s through joint public and private action, these failed in Southern Spain. Finally, because Southern owners were unwilling to finance even such secondary irrigation works as the levelling of the earth or the construction of conduits, they often did not use already existing facilities. A particularly notorious instance of this kind occurred in the Guadalete valley in Cadiz after the construction of the Guadalcacín Dam in 1910. Every government from that date until 1952 failed in its efforts to get the large local owners, most of whom were not nobles, to bring water from the principal canal to their farms. The impasse ended only when the Franco government itself undertook the building of the necessary secondary works. For forty-two years a major state investment remained useless for want of the type of private cooperation which during the same period was transforming the economy of Aragon.*[26]

As a total contrast to this indifference of big landowners, Malefakis describes the technical and juridical sophistication of the small-scale irrigation systems of eastern Spain:

> *Though they were attributed to the Moors, the Moors themselves regarded the irrigation works as an inheritance from a previous Christian civilisation; age-old, they represent a degree of social organisation and co-operation that compels admiration. Water belonged to the community and was sold with the land. The constant disputes as to its use in time of scarcity were regulated by a communal organisation . . . In later irrigation schemes water belonged to the capitalists who had built the system and was auctioned to the cultivators: hence the conflict, inconceivable in the older systems, between water-owners and cultivators that was typical of the nineteenth century.*[27]

In the *huertas* of Aragon and elsewhere the water belongs to the farmers and growers, large and small, through whose land it passes, and each water-user belongs to a *comunidad de regantes* (association of water-users) which elects a *sindico*, and the combination of *sindicos* from each zone constitutes the Water Tribunal. The most famous of all, the *Tribunal de las Aguas*, has become a well-known tourist attraction in Valencia. It meets every Thursday of the year at noon, sitting in a circle outside one of the doors of the ancient cathedral. Eight farmers, each representing one of the networks of canals, channels and drains of the 2,300 acres (930 ha) of the *huerta*, meet to judge infringements or disputes or to apportion rations of water in periods of scarcity. No lawyers and none of the state's laws are involved. Proceedings are verbal and are not recorded. Fines are sometimes imposed and are always paid. The tribunal is said to have existed continuously since its foundation by the Moors in 960 AD.[28]

In 1936 a conspiracy of army generals aimed to overthrow the government of the Popular Front in Madrid. Three governments in one day resolved to submit to the insurgent generals but were forestalled by a popular revolution determined both to resist the generals and to achieve the collectivization of agriculture and industry. The revolution was eventually suppressed, not by Franco's eventual victory, but by the government's submission to the foreign policy of Stalin's Russia. A variety of accounts of collectivized agriculture exist.[29] One eye-witness report, by Gaston Leval, describes the effect on the water tribunals in upper Aragon. He mentions the significance of bodies like the Valencia tribunal in settling disputes amicably, and goes on to say:

> *But such disputes disappear when men no longer have to compete and fight each other to exist, or when the will to acquire wealth for oneself is no longer uppermost. In the region of Fraga, fifteen* comunidades de regantes *covering the land in five villages, disbanded. The morality of solidarity produced that miracle. The old practice was replaced by a single collectivist administration, which coordinated the distribution of water everywhere, and which was proposing to improve the catchment basin and use of the rivers, especially of the Cinca river, by public works which none of the villages could have carried out individually.*[30]

Leval was there at the time, but, sixty years later, I think he was wrong to welcome the disbandment of the water tribunals. This is not because neither he nor the peasants could have known that the collectives would be suppressed, before Franco's victory, by the communists, but because the *comunidades de regantes* were themselves an achievement of popular self-organization.

In the early, punitive years of the Franco regime they were restored, but Spain subsequently changed beyond recognition. One historian of the post-war decades, Stanley Payne, explains how

> *The real Spanish revolution was not the defeated struggle of 1936–39 but the social and cultural transformation wrought by the industrialisation of the 1960s and 1970s . . . Even in some of the prosperous agrarian areas of the north-east, the agrarian population declined because of the inherent unattractiveness of a rural lifestyle in an urban, hedonistic, consumer-oriented society.*[31]

And Fred Pearce, after praising the survival of the Valencia tribunal, goes on to explain, gloomily, that

> *Out on the plains, between the large canal systems, there are dozens of smaller-scale village networks . . . and only in the past thirty years has the system crumbled. At Ahin (which is Arabic for 'spring'), for instance, three large cisterns collected spring water through each winter for a thousand years. Each June for a few weeks, more than a hundred farmers opened the cisterns and water flowed along a kilometre-long canal to irrigate crops of wheat, maize, beans, cherries and almonds. But since the 1960s, many farmers have left the village, the cisterns and canals have silted up, taps have gone unopened and, since about 1970, no records of the irrigation system have been kept up.*[32]

In 1990 Spain experienced the beginnings of a long period of prolonged drought. The traditional *comunidades de regantes*, lovingly described by Costa, had elaborate mechanisms for regulating the opening of sluice-gates for days and hours according to the quantity of water available and according to the crop. He explained how, in times of extreme drought, the *sindaco* of each channel allotted water with a view to saving as much of the crop as possible, sometimes giving preference to those crops which were most demanding, sometimes denying irrigation to those which could resist water shortage longest, sometimes giving water to half a field. Misfortunes were shared. But in the Spain of the 1990s the government in Madrid approved the National Hydrological Plan which would involve the building of 272 new dams and vast aqueducts moving water from the wet north to the dry south.

The proposals have brought a familiar conflict between the engineering approach and the conservationist lobby. Juan Lopez de Urralde, campaign director for Greenpeace, observed that

> *The Ministry has an engineers' lobby which has acquired a momentum of its own, and is continuing to apply a policy that goes back to the days of Franco, consisting essentially of huge construction projects. Greenpeace's counter-proposal is to start by looking at demand.*[33]

And Antonio Labajo of the National Metrological Institute explained to John Hooper that modern Spain has two different water problems, one meteorological and the other cultural. For,

> *Encouraged by a surge in prosperity since the early sixties, the Spanish have chosen to forget the fact that they live in a semi-arid country, large swathes of which are prone to periodic, lengthy droughts. They have built golf courses for their tourists, and swimming pools for themselves. In towns and cities the length and breadth of the country, they have created lawns and gardens that require daily watering. But, above all, they have allowed – and often encouraged – their farmers to diversify out of traditional drought-resistant produce such as figs and olives into water-hungry crops like rice and strawberries. The result is that Spain today is the world's fourth highest per capita consumer of water, after the US, Canada and Russia.*[34]

The water history of Spain demonstrates that the tragedy of the commons is not the one identified by Garrett Hardin. Communal control developed an elaborate and sophisticated system of fair shares for all. The private property recommended by Hardin resulted in the selfish individualism that he thought was inevitable with common access, or in the lofty indifference of the big landowners.

But the modern world has brought other factors into the equation: the assumption of a global market among producers, the assumption of a right to squander resources among consumers, and the assumption that the centralized state, with its control of revenue, its command of resources and expertise, and its power to enforce policy upon its citizens, should be the arbiter of the management of natural territorial and aquatic resources. We have to turn now to the fascination exerted by hydraulic engineering on people with power.

Notes

1. Joaquín Costa, *Colectivismo Agrario en Espana* (Madrid: Imprenta de San Francisco de Sales, 1898; Buenos Aires: Editorial Américalee, 1943). p. 403.

2. Garrett Hardin, 'The Tragedy of the Commons', *Science*, vol. 162, 13 December 1968; pp. 1243–8.

3. Green Alliance, *The Little Green Book* (London: Wildwood House, 1979), p. 17.

4. Hardin, *op. cit*, p. 1247, Garrett Hardin, *Exploring New Ethics for Survival* (New York: Viking, 1972).

5. Kirkpatrick Sale, *Human Scale* (London: Secker & Warburg, 1980), p. 334.

6. *Ibid.*, p. 334.

7. *Ibid.*, pp. 334–5.

8. *Ibid.*, p. 335.

9. Peter Marshall, *Nature's Web: An Exploration of Ecological Thinking* (London: Simon & Schuster, 1992), pp. 458-9.

10. *Ibid.*, pp. 340–1.

11. Hardin, *op. cit*, p. 1245.

12. *Ibid.*, p. 1245.

13. *Ibid.*, p. 1245.

14. K. D. M. Snell, *Annals of the Labouring Poor: Social Change and Agrarian England 1660–1900* (Cambridge University Press; 1985) pp. 184–5.

15. Michael Thompson, Michael Warburton and Tom Hatley, *Uncertainty on a Himalayan Scale* (London: Ethnographia/ Milton Ash Editions, 1986), p. 10.

16. Hardin, *op. cit*, p. 1247.

17. Robin Clarke, *Water: The International Crisis* (London: Earthscan Publications, 1991), p. 160.

18. Raymond Carr, *Modern Spain 1875–1980* (Oxford University Press, 1981), p. 17.

19. Ruth Way, *A Geography of Spain and Portugal* (London: Methuen, 1962), p. 105.

20. George Cheyne, *A Bibliographical Study of the Writings of Joaquín Costa* (London: Tamesis Books, 1972).

21. Costa, *op. cit* (n. 1, above).

22. Way, *op. cit*, p. 102.

23. Carr, *op. cit*, p. 21.

24. Edward E. Malefakis, *Agrarian Reform and Peasant Revolution in Spain: Origins of the Civil War* (New York: Yale University Press, 1970).

25. Carr, *op. cit*, p. 21.

26. Malefakis, *op. cit*, p. 80.

27. *Ibid.*, p. 19.

28. Victor Fairén-Guillén, *El Tribunal de las Aguas de Valencia y su Proceso* (Valencia: Caja de Aborros, 1988).

29. Burnett Bolloten, *The Spanish Revolution* (Chapel Hill: University of North Carolina Press, 1979); Pierre Broué and Emile Témine, *The Revolution and the Civil War in Spain* (London: Faber & Faber, 1972).

30. Gaston Leval, *Collectives in the Spanish Revolution* (London: Freedom Press, 1975), p. 112.

31. Stanley G. Payne, *The Franco Regime 1936–1975* (University of Wisconsin Press, 1983), p. 107.

32. Fred Pearce, *The Dammed: Rivers, Dams and the Coming World Water Crisis* (London: The Bodley Head, 1992), p. 23.

33. John Hooper, 'The drain on Spain', *Guardian*, 14 June 1995.

34. *Ibid.*

CHAPTER 3

Hydraulic Societies and Regional Hopes

Actually so much water is moved around California by so many different agencies that maybe only the movers themselves know on any given day whose water is where, but to get a general picture it is necessary only to remember that Los Angeles moves some of it, San Francisco moves some of it, the Bureau of Reclamation Central Valley Project moves some of it and the California State Water Project moves most of the rest of it, moves a vast amount of it, moves more water farther than has ever been moved anywhere. They collect this water up in the granite keeps of the Sierra Nevada and they store roughly a trillion gallons of it behind the Oroville Dam and every morning, down at the Project's headquarters in Sacramento, they decide how much of their water they want to move the next day. They make this morning decision according to supply and demand, which is simple in theory but rather more complicated in practice. In theory each of the Project's five field divisions places a call to headquarters before nine a.m. and tell the dispatchers how much water is needed by its local water contractors, who have in turn based their morning estimates on orders from growers and other big users. A schedule is made. The gates open and close according to schedule.

Joan Didion, *The White Album* [1]

In her absorbing essay on 'Holy Water', Joan Didion reminds us that her West Coast region of the United States is totally dependent on the vast network of engineering projects she describes with awe. She recalls how within her own memory the typical

Californian summer was characterized by 'the coughing in the pipes that meant the well was dry', and its winters by 'all-night watches on rivers about to crest, by sand-bagging, by dynamite on the levees and flooding on the first floor'.[2] She regards the apparent ease of Californian life as an illusion, but sees that, in practice, demands are met, even though there might be a sudden need to hold 'large flows of water for power production, or to flush out encroaching salinity in the Sacramento–San Joaquin Delta, the most ecologically sensitive point on the system'.[3] In practice, too, a sudden fall of rain might obviate the need for a delivery of water which has been working its way through the system for two days.

Who is to manage such an endlessly complex system? Joan Didion observes that it requires 'prodigious coordination, precision, and the best efforts of several human minds and that of a Univac 418'.[4] Her conclusion is that 'Water is important to people who do not have it, and the same is true of control'.[5]

Control of water is inevitably control of life and livelihood, and this thought occurred to a Marxist scholar, Karl Wittfogel, who devoted a great deal of scholarship to the study of those ancient civilizations that Marx described as 'oriental despotism'. He was a member of the 'Frankfurt School' of historians and thought his research threw light on the fate of the Soviet Union. He was an early victim of the Nazis and was fortunate in being able to emigrate to America, where he became a busy anti-communist. There, his book *Oriental Despotism*, which appeared in 1957, was regarded as a Cold War weapon directed at Stalin's Russia.[6] Its title was useful, but he had concluded that the designations 'hydraulic society' and 'hydraulic civilization' better described his theme.

His book had several arguments, of which the most important was that ancient empires like those of China and Egypt were built upon central control, through a vast bureaucracy, of the waters of great rivers, determining the livelihood of those inhabitants who depended on irrigated farming.

He added that

> *In recent years man has uncovered the productive energies of electricity. Again he is turning his attention to falling water. But even when the engineer of the 20th century erects his power plant on the very spot that previously supported a textile mill, he actualises new forces in the old*

setting . . . What is true for the industrial scene is equally true for the agricultural landscape. The hydraulic potential of the earth's water-deficient regions is actualised only under specific historical circumstances. Primitive man has known water-deficient regions since time immemorial; but while he depended on gathering, hunting, and fishing, he had little need for planned water control. Only after he learned to utilise the reproductive processes of plant life did he begin to appreciate the agricultural possibilities of dry areas, which contained sources of water supply other than on-the-spot rainfall . . . only then did the opportunity arise for despotic patterns of government and society.[7]

His generalizations were more modest than his critics lead us to believe. For example, he stressed that many such societies, like those of the Po Plain in Italy, or that of the Netherlands, did not generate 'despotic methods of statecraft'; but the irony of the impact of his book was that, in the form in which it was summarized for nascent politicians in American and British universities, the afterwash of his ideas formed the basis for the technocratic ideology that claims that central government knows best, and that short-sighted displaced local populations are simply oblivious of the fact that their extinction or removal is a sacrifice for the general good.

The parallels he drew with Stalinist Russia's centralized industrialization do not concern us any more, but for a great many readers the message he left behind was that centralized planning of water resources was a law of nature. And in a strange way, Wittfogel's convoluted theories about oriental despotism and hydraulic societies in antiquity are being realized in the modern world, thanks to sophisticated modern civil engineering techniques, and the lure of vast international contracts which can only be realized at the level of central governments, international free-floating capital and supra-national governmental organizations. Real life is providing ample examples of the power of oriental despotism.

Trying to bridge the gap between theories and real life, Ernest Gellner commented that

It is not clear that Wittfogel was right about irrigation; Edmund Leach on Sri Lanka, and Robert Fernes on the marshes of southern

> *Iraq, claim that complicated water-works are better run by decentralised than by centralised politics. But whether or not he was right about agrarian irrigation systems, the insinuated modern analogy proved unquestionably wrong; a centralised despotism is not a good way of managing industrial growth. The Soviet Union has demonstrated this with a conclusiveness seldom found in these matters.*[8]

But even Gellner's optimism about the lessons of history was outflanked by the determination of modern despots to build monuments that leave a permanent mark on their countries through manipulating water. Taking his example of Sri Lanka, Fred Pearce reported in 1992 that the disruption caused by the Mahaweli dams and plantation project had forcibly moved up to a million people and was seen as 'a prime cause of the unrest that saw tens of thousands dead as the Tamil Tigers fought the government's forces in the jungles of eastern Sri Lanka from the late 1980s onwards'.[9]

And taking Gellner's example of the Marsh Arabs of southern Iraq, it was reported in 1993 that the President

> *has secretly extended his plan to drain Iraq's southern marshlands, putting his engineers to work east of the Tigris river for the first time and warning that anyone who opposes them risks death . . . The area affected by the new drainage scheme – the Mosharah marshes enclosed by the towns of Amara, Mosharah and Kahla – is the most heavily populated part of the marsh region. Many of its estimated 100,000 inhabitants are already fleeing.*[10]

An example which both supports the Wittfogel thesis, but reveals another side – that of its absorption into the folk culture of local water management, and one that is in operation to this day – comes from ancient China. Robin Clarke links it to our current preoccupation with large-scale water engineering:

> *In 250 BC a regional governor in Szechwan, Li Ping, and his son Er Wang, devised a scheme to tame the floods of the Min River caused by melting snow in Tibet. More than a thousand kilometres of canals were constructed which provided irrigation for 200,000 hectares . . . The main diversion canal is separated from the Min River by a*

structure known as the fish snout, which is regularly damaged by the floods. But every year, when the river is low, the canals are cleaned and the fish snout repaired. Flow in the canals is controlled by simple and temporary structures of sausages of rock held together by twisted bamboos. This primitive, yet appropriate system – for it works – has proved capable of controlling an immensely complex hydrological system for more than two millennia.[11]

Clarke follows his description with the reflection that

In many senses, the Min River control system exemplifies a quality totally lacking in today's grandiose water schemes: the combination of large-scale, overall control with small-scale, simple structures. No one item in the Min River scheme is large; even the fish snout that first divides the river is only tens of metres long. Nor is the technology complex. Since it can be easily repaired, it does not have to be built to withstand exceptional conditions. Sedimentation is simply controlled by digging out sections of the canal, and then using the rich nutrients obtained to boost crop production. Such ideas would be totally impractical for a modern reservoir, covering hundreds of square kilometres, and designed specifically never to dry out.[12]

This is very significant observation. But where was the element of overall control, except in the oriental despotism of two thousand years ago? If you consider the endless cataclysms of China's history in the past hundred years, it is impossible to imagine continuity of control of the Min River from some central bureaucracy. The elaborate system maintained itself because it was in the obvious interests of innumerable local communities to undertake the seasonal tasks with the means available to them. Clarke was right in stressing the contrast with today's huge engineering projects.

Kate de Selincourt tried to dramatize this difference by adding up the numbers of people who stand to lose their homes, their livelihood and their land, even their cultural identity, through actual and projected giant dams. The figure she reached was three million:

The Kedung Ombo dam in Indonesia displaced 25,000 people, the Akasombo dam across the Volta in Ghana displaced 80,000. On the Zambesi in southern Africa, Cabora Bassa displaced 25,000 and Kariba 50,000. Dams are proposed along the Mekong from Vietnam up through Cambodia, Lao PDR and Thailand. Three, in Lao PDR, will displace 142,000 people between them. In China, the Xiao Langdi dam proposed for the Yellow River would displace 140,000 people, while the Three Gorges project would displace at least 1.1 million.[13]

The appalling devastation of human lives and livelihoods that these grand projects entail makes hydraulic engineering on such a vast scale as tragic and futile as war, and the record in delivering water and power and alleviating water shortages or eliminating flooding further down the valleys is so poor, and so surrounded by new problems, that we have to wonder why today's oriental despotisms dream of embarking on such projects, since a thousand little local projects would be so much more beneficial. The answer must lie in the achievements of engineers, both civil and military, in the period between the two world wars.

The incredible network of water management that Joan Didion celebrated was the result of vast projects like the Hoover Dam, which, as Fred Pearce put it, transformed the US government's Bureau of Reclamation into 'a mass producer of large dams' originally funded by the Federal government, but eventually functioning as autonomous enterprises, since

> *As long ago as 1906 the Bureau had persuaded Congress to allow it to keep income from the sale of electricity generated at its dams for ploughing back into irrigation schemes, and with the Hoover Dam, it had created a huge money-making machine for itself. Soon the money-making became an end in itself. Early ideals that electricity should be preferentially offered to cooperatives and municipal authorities gave way in the 1940s to a series of massive contracts with private utilities.*[14]

The Hoover Dam was followed by the Parker Dam, the Imperial Dam and the Oroville Dam, and in the North-West by the Grand Coulee Dam, completed in 1942, and more than twice the size of the Hoover Dam, which itself had been seen in the 1930s as the greatest engineering achievement of the twentieth century. But the venture which was seen throughout the world as the supreme example of the use of water engineering to bring new hope to a whole region, was the Tennessee Valley Authority, which we have to examine in detail.

The Tennessee River in the east of the United States is shaped like an arm with the elbow bent. The outstretched fingers are the five large tributaries rising in the Blue Ridge mountains of West Virginia, in the Great Smokies of North Carolina and the southern Appalachians. The forearm is the upper Tennessee, the elbow is at Muscle Shoals (the rapids where the river turns north). The upper arm is the lower Tennessee river, and the shoulder is the westward curve of the river in Kentucky, where it joins the Ohio river, which in turn flows south into the Mississippi.

To a European, even a physical description sounds like a catalogue of American folk music, and to a regional geographer the natural boundaries of the region are not the seven state-lines or political boundaries crossed by the river and its tributaries, but their watersheds. The whole vast drainage basin comprises 42,000 square miles, about the size of England.

By 1933, as Herbert Agar put it, the area's livelihood had drained away with its topsoil,

> *The balance of nature was upset. A land of abundance became a land of rural slums, lacking even the sad companionship which city slums afford. Men's minds and characters decayed, as faith diminished and malnutrition became customary. Racial hatreds flourished, because the 'poor white' had nothing left to comfort him except an evil sense of superiority to the Negro.*[15]

The isolated valleys in the region were occupied by subsistence farmers attempting to grow cash crops like cotton, tobacco or maize, and as the yields of the valley fields diminished, they cut down the trees, burned off the vegetation and ploughed the hill slopes, moving further and further up to the mountain sides. The heavy rainfall, the failure to replace the land's fertility, and the removal of the forest cover, allowed the soil to wash away into the rivers, so that, in Julian Huxley's words, 'in the heart of the most modern of countries you could find shifting cultivation of the type usually associated with primitive African tribes'.[16] In this 650-mile long river basin the eastern Appalachian area was the poorest part of the poorest region in the whole of the United States.

There are two possible explanations for the programme of rescue that followed: anarchist and governmental. Our foremost regional geographer, Prof. Peter Hall, traces what he calls 'the anarchist roots of the planning movement' through a tradition which flows through the founding fathers of French geography, Paul Vidal de la Blache (1845–1918) and Élisée Reclus (1830–1905), through the Russian geographer Peter Kropotkin (1842–1921) and the Scottish sociologist Patrick Geddes (1854–1932).

> *For Vidal and his followers, as for Geddes, regional study gave understanding of an 'active, experienced environment' which 'was the motor power of human development; the almost sensual reciprocity between men and women and their surroundings was the seat of comprehensible liberty and the mainspring of cultural evolution', which were being attacked and eroded by the centralised nation-state and by large-scale machine industry . . . For Geddes, as for Vidal, the region was more than an object of survey; it was to provide the basis for the total reconstruction of social and political life.*[17]

Peter Hall explained that 'one of the most crucial insights that Geddes borrowed from Kropotkin' was the impact of technical change, the prospect that 'new sources of power, hydraulic and especially electric, meant that a big central unit of power was no longer needed; industries that depended chiefly on skilled labour had no economies of scale; observably the newer industries tended to be small in scale'.[18] He went on to cite key passages from Kropotkin's book *Fields, Factories and Workshops*, explaining that the big industrial concentrations of the nineteenth century represented pure historical inertia, a legacy destined to die out, as indeed they have, to the immense cost of their populations, by the end of the twentieth century. Kropotkin had argued that

> *There is absolutely no reason why these and like anomalies should persist. The industries must be scattered all over the world; and the scattering of industries amidst the civilised nations will be necessarily followed by a further scattering of factories over the territories of each nation . . .*
>
> *This scattering of industries over the country – so as to bring the factory amidst the fields, to make agriculture derive all those profits*

which it always finds in being combined with industry . . . and to produce a combination of industrial with agricultural work – is surely the next step . . . This step is imposed by the very necessity of producing for the producers themselves; it is imposed by the necessity for each healthy man and woman to spend a part of their life in manual work in the free air.[19]

In Hall's view 'this was one of the most crucial insights that Geddes borrowed from Kropotkin', and he notes that Geddes 'directly followed Kropotkin' in anticipating an era in which we would 'apply our constructive skill, our vital energies, towards the public conservation instead of the private dissipation of resources, and towards the evolution instead of the destruction of the lives of others'.[20] And in the specialist world of American advocates of these ideas, the Regional Planning Association of America (RPAA) was started in March 1923 as a result of chance meetings of Geddes' most influential disciple, Lewis Mumford, with the architects Clarence Stein and Henry Wright and the proposer of an Appalachian Trail, Benton MacKaye. In June that year, Patrick Geddes visited New York, and the RPAA adopted a programme which included the creation of garden cities within a regional plan, development of regional projects, 'and surveys of key areas, notably the Tennessee Valley basin'.[21]

The people influenced by this approach were provided with the opportunity to apply it by a quite different, governmental bunch of enthusiasts for the Tennessee Valley. Theodore Roosevelt, who as vice-president succeeded to the presidency of the USA in 1901 and was re-elected in 1904, was famous both for his bellicose, imperialist policies abroad and for his concern with the conservation of natural resources at home. It was his ideas that led to the establishment of both military and civilian bodies given the task of reporting on the problems and potentialities of the regions of the United States.

There was very little industry in the Tennessee Valley, a region which, like most of the South, had been impoverished since the Civil War. But during the First World War the Federal government had built nitrate plants for the production of explosives and had begun a dam to supply them with electric power. The dam was completed after the war, and for years was the subject of political controversy. The Republican governments of Coolidge and Hoover proposed to sell the plants and dam to Henry Ford, but the sale was blocked by Senator George Norris, whose bills for federal control were, in turn, vetoed by the presidents. Meanwhile two army engineers, General Brown and Major Watson, had, by 1930, published a comprehensive development plan in a 734-page report called *The Tennessee River and Its Tributaries*. Norris, as a result of this and the studies made by the RPAA, the National Conservation Commission and the Inland Waterways Commission, was ready for a change in the political climate.

In 1932, speaking at the RPAA conference, Franklin D. Roosevelt commended the idea as an example of regional planning, 'but this, like so much of his language, was "a phrase so loose and imprecise that it could fit almost any program, and yet so elusive that it involved few specific commitments"'.[22] A few months later, with Roosevelt's inauguration as president, the TVA was created on 18 May 1933 as one of the first planks of his New Deal policy to lift America out of depression. To its board he appointed three members who would make, as Peter Hall explains, 'a totally, explosively incompatible mixture'.[23] The chairman was A. E. Morgan, President of Antioch College. He 'had much in common with the early utopian communitarians' and 'saw the job as his life's opportunity to realise his personal vision of a new physical and cultural environment: a vision he believed FDR to share'.[24] He hoped that the new town of Norris would follow the Kropotkinian vision of a place where the inhabitants would combine agriculture and industry. He was soon in dispute with his colleagues, and was dismissed by Roosevelt in 1938.[25]

The second member was David Lilienthal, described by Peter Hall as 'an immensely ambitious, driving young man with the reputation of stealing any show he joined'.[26] He was the author of the official account of the TVA,[27] and was in charge of power generation and the vast engineering works associated with water control. They were, in fact, a vast technical achievement. By the end of its first 20 years, the TVA had built twenty dams across the river and its tributaries, and these together with those which existed before 1933 form a system of 28 reservoirs, controlling the flow of water by a defence-in-depth system of flood control which, along the Tennessee River and the lower Ohio and Mississippi, brought annual savings of many millions of dollars worth of otherwise unavoidable damage. The locks and reservoirs form 630 miles of navigable canal from Paducah to Knoxville. In 1933, transport on the Tennessee River, ruined for navigation by the soil washed down from the hillsides, was less than 33 million ton-miles. By 1951 it was over 589 million ton-miles. In terms of power-generation, the TVA was over-successful. When the Authority began its dam-building activities it was prophesised that there could be no possible use for the power generated. The electricity-producing capacity of the region in 1933 was under 815,000 kilowatts. By the mid-1950s it was 4 million kilowatts, but already in 1951 power had to be borrowed from other systems, power-cuts and load-shedding had to be introduced. The Authority began building five steam-generating power stations intended to burn 8 million tons of coal annually. Subsequently it added nuclear power stations. A commentary in 1954 remarked that,

> *If you can conceive of an ecology of industry, then the Tennessee Valley has already moved from the unbalance of wasted resources to the over-exploitation of resources. The difference between power generated from water and that generated from burning coal is obvious. One is inexhaustible and the other uses material which cannot be renewed. When coal has to be imported by the largest producer of hydro-electric power in America, something has gone wrong.*[28]

Something had indeed gone wrong, but before revealing it, we should turn to the third board member in the original TVA triumvirate. This was Harcourt A. Morgan, no relation of the chairman, and characterized by Peter Hall as

> *President of the University of Tennessee, representative of the conservative agrarian interests at Vanderbilt, obsessed with the idea of rural extension services and in particular with a scheme for a phosphate fertilizer programme. He readily made common cause with Lilienthal . . . But of regional planning – especially that radical variant espoused by the RPAA – there was an imperceptible residue: community development, health and educational services got a minuscule sliver of the total budget.*[29]

However, Harcourt Morgan's achievements were real enough. William E. Cole reported in 1950 that

> The device selected by TVA and the state agricultural colleges to bridge the gap between scientific research and practical application is the test-demonstration farm. We have thousands of test-demonstration farms in the Tennessee Valley; about one out of every ten farms. These are normal going farms owned and operated by typical farmers. Their purpose is two-fold: first, to test the effectiveness of new kinds of fertilisers being developed by TVA; second, to demonstrate the proper use of the proper kinds of fertilisers in soil and water-conserving systems of farming. The test-demonstration farmers are selected by their neighbours, not appointed by the Government.[30]

In the 20 years after 1933, 200 million trees were planted in the valley. In 1933 every third employee of a hosiery firm in Decatur was found to be suffering from malaria. Ten years later the figure had dropped below 1 per cent. By 1951 the TVA was able to report that

> A million acres of land has been terraced, another million acres has been shifted from row to close-growing crops, and improved permanent pastures have been increased by 800,000 acres. More forage and pasture have increased the number of livestock and encouraged dairying. More diversified farming is making possible more efficient use of farm labour. Most of the farms in the Valley now have electricity. The region today is much more capable of producing food and fibre than it was before the war. But it is far from the level of fertility and productiveness justified by its soil base, its favourable growing climate, its ample rainfall, its growing scientific and technical knowledge, the energy and initiative of its people, and their basic love of the land.[31]

The Tennessee Valley Authority exists to this day, with a staff of 24,000 people divided between an Office of Coal Gasification, an Office of Agricultural and Chemical Development, an Office of Power, an Office of Engineering Design and Construction, an Office of Economic and Community Development and an Office of Natural Resources. Its main source of revenue is the sale of electricity to the 160-odd municipal utilities and corporations that have contracted with the TVA as a wholesale supplier. For an assessment of its significance I turn to William Ashworth's *Encyclopedia of Environmental Studies*:

> Controversial from the beginning, TVA survived a series of court challenges in the late 1930s that sought to have it declared unconstitutional and several similar challenges following World War II seeking to limit its scope to water resource development. Though from an environmental standpoint it is easy to argue with the agency's full-development bias, it is also true that TVA has provided an extraordinarily successful model for integrated resource management based on watershed boundaries rather than on political borders.[32]

However, there is one more political truth about the TVA that explains its preoccupation with power generation, including nuclear power plants. When the Authority was created, the military programme of nitrate production for explosives was altered to the production of nitrate fertilizers. But in the Second World War the Atomic Energy Commission selected the Tennessee Valley, with its ample cheap electricity, as the place for the production of plutonium for the atomic bomb. By 1944, the TVA 'was already the second-largest producer of power in the United States, generating half as much as the entire national production in 1941'.[33] Peter Hall comments on the irony that 'the one element that Roosevelt had removed from the TVA prescription, munitions production, was now driving the economic development of the Valley'.[34] And after the war, Senator Tom Stewart explained that 'one of the major reasons for locating the atomic bomb project in Tennessee is because there's a greater percentage of native-born American, patriotic labour in Tennessee and the South'.[35] Thousands of people worked night and day in the largest users of TVA power, the plants at Oak Ridge and Paducah, to produce the bombs that fell on Hiroshima and Nagasaki, without the slightest knowledge of what they were making.[36] Historians record that the work was done at fever pitch so that the bombs could be delivered before the rulers of Japan could accept the terms of unconditional surrender. Writing forty years ago about the deformation of the original purposes of the TVA, Peter de Mendelssohn remarked that 'One factor, of course, could transform and "normalise" this situation overnight – if Paducah and Oak Ridge were to shut down or go away. But this is, admittedly, the least likely of all developments.'[37] It was, and is.

The TVA story is full of contradictions. It was a creative harnessing of water power, the realization of the dreams of the pioneers of regional geography. It was administered in a way which may not have been quite so decentralized as propagandists hoped, but was very far from Wittfogel's image of 'oriental despotism' as the precondition for hydraulically-based societies. As we have seen, its policies were in the end determined by the differing interests of its board members. It accomplished a great deal in changing the prospects for the most poverty-stricken region of the United States.

But it was only accidents that made it possible. The need had been pressed for years both by the regional planners and the US army engineers. The chance that the revitalization of the Tennessee Valley was a visible symbol for Roosevelt's New Deal programme for lifting America out of the Great Depression, enabled him to override both the universal faith in private enterprise and the conflicting interests of the state governments. And it was only the accident that the US government's wartime development of nuclear weapons needed both an abundant source of power and a docile labour force that made the TVA financially successful. At the same time this very chance distorted hopes for an ecologically sound pattern of power generation.

These are the awkward paradoxes of the TVA. But it also left behind a worldwide assumption that large-scale water engineering was automatically a social good.

Notes

1. Joan Didion, *The White Album* (London: Weidenfeld & Nicolson, 1979), p. 60.
2. *Ibid.*, p. 59.
3. *Ibid.*, p. 60.
4. *Ibid.*, p. 61.
5. *Ibid.*, p. 64.
6. Karl August Wittfogel, *Oriental Despotism* (New Haven: Yale University Press, 1957).
7. *Ibid.*, p. 12.
8. Ernest Gellner, 'The last Marxists', *Times Literary Supplement*, 23 September 1994, p. 3.
9. Fred Pearce, *The Dammed: Rivers, Dams and the Coming World Water Crisis* (London: The Bodley Head, 1992), p. 155.
10. Julia Flint, 'Shi'ites flee new marsh drainage', *Guardian*, 13 November 1993.
11. Robin Clarke, *Water: The International Crisis* (London: Earthscan Publications, 1991), p. 141.
12. *Ibid.*, p. 141.
13. Kate de Selincourt, 'The dammed', *New Statesman & Society*, 12 March 1993.
14. Pearce, *op. cit*, p. 93.
15. Herbert Agar, cited in Henry Billings, *The Power and the Valley* (London: Hart-Davis, 1954), p. 21.
16. Julian Huxley, *TVA: Adventure in Planning* (London: Architectural Press, 1943), p. 9.
17. Peter Hall, *Cities of Tomorrow: An Intellectual History of Urban Planning and Design in the Twentieth Century* (Oxford: Basil Blackwell, 1988) p. 142.
18. *Ibid.*, p. 144–5.
19. Peter Kropotkin, *Fields, Factories and Workshops* (London: Hutchinson, 1899; updated edition edited by Colin Ward, London: Allen & Unwin, 1974; London: Freedom Press, 1984), pp. 157-8.
20. Hall, *op. cit*, p. 148.
21. *Ibid.*, p. 148.
22. *Ibid.*, p. 161, citing P. K. Conklin and E. C. Hargrove (eds), *TVA: Fifty Years of Grass-Roots Bureaucracy* (Urbana: University of Illinois Press, 1983).
23. *Ibid.*, p. 161.
24. *Ibid.*, p. 163.
25. For A. E. Morgan's account, see A. E. Morgan, *The Making of TVA* (Buffalo: Prometheus Books 1974). For a discussion of the issues involved, see T. K. McCraw, *Morgan vs. Lilienthal: The Feud within the TVA* (Chicago: Loyola University Press, 1970).
26. Hall, *op. cit*, p. 163.
27. David E. Lilienthal, *TVA: Democracy on the March* (Harmondsworth: Penguin Books, 1944).
28. Colin Ward, 'Second thoughts on TVA', *Freedom*, 11 September 1954, p. 4.
29. Hall, *op. cit*, p. 162.
30. William E. Cole, addressing the US Scientific Conference on Conservation and Utilisation of Resources, Washington 1950.
31. *Annual Report of the Tennessee Valley Authority* (Knoxville: Tennessee, 1951).
32. William Ashworth, *The Encyclopedia of Environmental Studies* (New York/Oxford: Facts on File, 1991), p. 386.
33. Hall, *op. cit*, p. 163.
34. *Ibid.*, p. 163.
35. Quoted in Mortimer Slaiman, 'The atom bomb, Oak Ridge and the South', *New Leader*, USA, 24 May 1947.
36. See C. Allardice and E. R. Trapnell, *The Atomic Energy Commission* (New York: Praeger, 1974).
37. Peter de Mendelssohn, in *New Statesman*, 23 May 1953.

CHAPTER 4

The Lure of the Dam

> *Everywhere large water projects are both the consequence of and the justification for authoritarian government. From India to Turkey to Paraguay, soldiers stand guard over dam construction sites and herd refugees to their new homes. It is no surprise that one of the USA's great dam-building organisations is the US Army's Corps of Engineers. Or that Stalin's secret police supervised construction of his dams and canals. Or that soliders such as Nasser and Gadaffi and the military commanders of South America have been so prominent in the promotion of large dams. The urge, and frequently the language, of domination and rule runs through their world. It is a long way from the dreams of the pioneers of the 1930s who believed that dams could help bring peace, prosperity and social justice.*
>
> **Fred Pearce, *The Dammed*** [1]

In 1920, Lenin declared that 'Communism is Soviet power plus the electrification of the whole country', and this led his successors to embark on a series of vast-hydro-electric projects, using forced labour, from the Dneiper Dam to the Rogun Dam (said to be the world's highest, at 325 metres). The TVA was more broadly conceived, since it was initially planned to combine power generation with flood control, a navigation system, soil and forest conservation, and the development of model farms, educational and health services. The architect Percival Goodman was captivated by the idea of the TVA in 1934, and remained convinced at the end of his life that it demonstrated 'that a plan, even when huge in size and scope, need not be a manifest danger. Yet we are right', he added, 'in being suspicious . . .' [2]

The well-publicized story of the TVA was seen in the Second World War as an exemplar for the peacetime mobilization of the huge engineering resources assembled for wartime purposes. Welsh nationalists, for example, believed that it was a model for the regeneration of their country's depressed economy,[3] and in Scotland it inspired the development of hydro-electricity as a public good.

Tom Johnston was a Scottish Labour Party member of parliament who in the wartime coalition government was, to his surprise, appointed as Secretary of State for Scotland. He had seen 'the legal obstructions which political and commercial interests in the United States had placed in the way of the Tennessee Valley Authority and the savage opposition which the Ontario Power Commission had, in its early stages, to suffer',[4] and exploited the wartime political consensus to set up a commission to report on the potential of a public, non-profit corporation to generate and transmit electricity from Highland water power. By 1943 he had achieved an Act of Parliament to institute the North of Scotland Hydro-Electric board. At the end of the war he declined office in the incoming Labour government in order to become its chairman. Peter Hennessey, the historian of the period, remarks that 'It was the nearest Britain has ever come to having a Tennessee Valley Authority, a utility with a heart. It commands fierce local loyalty to this day.'[5]

Johnston stressed that the Hydro Board had wider responsibilities than power generation: it was required by statute to be not only a generator of electricity, but a distributor of it over 74 per cent of the land area of Scotland, including experimental generation of wind power in the islands, and was 'enjoined to promote and encourage the economic and social welfare of the Highlands'.[6] Nevertheless, following the market values of the Thatcher government of the 1980s, far from Scottish wishes, it has now become Scottish Hydro-Electric plc, a private company.

The Hydro Board embarked upon very sophisticated, long-term, ways of coping with the changing demand for electric power which are more sensitive than those of fossil-fuel or nuclear generation. Thus a visitor to the Cruachan dam and power station in the West Highlands explains that:

construction of the dam, and more especially of the underground
..ies, tunnels, pipes and vast machine-hall hewn out of the
mountain's granite heart, are one of the greater feats of modern
engineering in Britain. The dam itself, although eye-catching, is not
one of the scheme's marvels: these are hidden below at a cost of
twenty-four million pounds. The Cruachan scheme was opened in
1965 as a pumped storage station, the second biggest of its kind in
the world (the largest is in Luxembourg). Its four generating turbines
have a capacity of 400 megawatts and an average yearly output of
450 million units (more than the nuclear power station at Hunterston
in Ayrshire) and they are reversible – they can drive the water back up
the mountain into the reservoir behind the dam. They do this by way
of two steep shafts, each 16 feet wide, which have been blasted
through solid rock. There is no way of storing electricity in great
amounts except by using it thus to rebuild water power . . . Energy
that would otherwise go to waste is stored in the Cruachan corrie as
water, available at a few minutes notice to meet any demand.[7]

It is easy to see that the sophisticated engineering expertise that can devise such a scheme is in itself a valuable commodity and that the people with this kind of experience are anxious to build upon it and to apply it wherever a government is willing to buy it and to invest in the enormous capital cost of large-scale water manipulation, not in sparsely populated areas like the Scottish Highlands (where in the early nineteenth century large numbers of the human population were forcibly evicted to make room for sheep), but in populous river valleys where the displaced inhabitants would gain nothing from huge engineering projects.

The Egyptian Nile was one of the exemplars of Wittfogel's theory of hydraulic civilizations, even though, as the anthropologist Harold Barclay stresses,

Canals associated with the Nile were primarily built for
transportation of stone for pyramid building and other public works or
for draining swamps. Egyptian sources given no indication whatsoever
of a role for irrigation canals in administration. If such a technology
were actually crucial for the creation of the Egyptian state one would
expect otherwise.[8]

In fact the despot who began the modern rearrangement of the

Nile was Mohammed Ali, the Ottoman empire's Viceroy of Egypt, manipulated by Napoleon and the French demand for raw cotton. He restored the ancient canal linking Alexandria with the Nile. Fred Pearce describes how

> The job took three months, and 20,000 forcibly conscripted labourers lost their lives in the process. Next, he sought to tame the Nile and the farmers who had captured its flood and planted crops in its silt for 7,000 years. He wanted to convert the farms of the Nile valley and delta to year-round irrigation, so that they could grow both a food and a cash crop each year. To do this, he once again press-ganged hundreds of thousands of peasants to dig canals to provide, for the first time, a permanent flow of water to the fields.[9]

By the 1880s, the British had replaced the French as controllers of the nominally Ottoman Egypt, and the celebrated water engineer William Willcocks was the builder of the first Aswan Dam, which yielded a much-increased crop of high quality cotton for the Lancashire industry every spring and more maize every summer to feed the ever-increasing population of Cairo.

> Nothing like it had been built before. Like a huge weir, it let the early silt-laden floodwaters from the Blue Nile pass over its top each summer. Then it captured up to one cubic kilometre of the clear water from the slightly later and smaller flood of the White Nile in a reservoir behind the dam. The reservoir's water was released gradually during the dry season, increasing the Nile's meagre flow.[10]

Fifty years after the first Aswan Dam was built, a bloodless military coup in Cairo finally ended foreign dominance of Egypt and its playboy puppet, King Farouk. Gamal Abdel Nasser became first prime minister and then president, and embarked upon the Aswan High Dam, a far greater venture intended to keep the level of the Nile constant all through the year, without flooding, and to provide innumerable ancillary benefits.

In the Cold War struggle for influence over the 'unaligned' governments, the United States and Britain were at first to provide the expertise and finance for the High Dam. When they withdrew, and after the British and French governments had made war on Egypt over Nasser's nationalization of the Suez Canal, the Soviet

Union stepped in to meet most of the cost of construction. Work began in 1960, and the dam was brought into operation in 1971, a year after Nasser's death.

Attitudes to the Aswan High Dam were determined by many factors. Other dictators resolved to emulate Nasser in leaving their permanent mark on the history of their countries. Western rulers, outraged by the Egyptian government's wish to show that it could do without them, were anxious to prove that it was an expensive failure. In the Middle East, Nasser's rivals for influence in the Arab world echoed the opinion of Mohammed Ali's grandson and successor a century earlier, that attempts to control the flow of the Nile were 'a colossal crime against the laws of nature'.[11]

However, after a quarter of a century, it becomes possible to see some of the long-term consequences. The absence of the annual flow of silt in the Delta has brought a need for a growing use of expensive chemical fertilizers, and an increasing vulnerability to erosion from the Mediterranean. At the same time the former annual flooding upstream used to wash away the build-up of natural salts, now left behind to increase the salt content of irrigated land. One of the blessings of the dam was hydro-electricity, in a country very short of fuels. But the build-up of silt trapped on the bed of Lake Nasser behind the dam is steadily reducing the generation capacity of the dam, while the lake itself is blamed for a dramatic increase in water-borne diseases. Meanwhile, upstream from the Egyptian Nile, other countries are developing their own plans for the White Nile and the Blue Nile. In 1990 Boutros Boutros-Ghali, then Egypt's Foreign Minister, reminded the Africa Water Summit Meeting in Cairo that 'The national security of Egypt, which is based on the water of the Nile, is in the hands of eight other African countries.'[12]

The more ruthless and unopposable any regime is, the grander are its adventures in water engineering. Thus the Soviet Union embarked on a series of vast irrigation projects in the rivers that feed the inland Aral Sea, bordered by Kazakhstan and Uzbekistan. The intention was to increase production of cotton and rice. The result has been a continual shrinking of the Sea which once provided 25,000 tonnes of fish a year and now yields none, as salinity has trebled, while over-irrigation has washed away the humus of the soil. To restore the sea, it was proposed to divert the flow of two Siberian rivers, but then the centralized government fell apart, leaving a collapsed ecology.[13]

And thus Colonel Gadaffi in Libya has committed his country to the largest civil engineering contract ever signed, in constructing the 'Great Man-made Rivers', pumping water from far beneath the Sahara in huge metal pipes to the coast. 'No nation', comments Fred Pearce, 'has ever mortgaged so much of its future on a single project', to create 'the most expensively watered fields in the world'.[14] The Libyan adventure has the saving grace of not disturbing the livelihoods of existing human communities. In the destruction of the Aral Sea, half the population of the city of Aralsk has fled. But the proposed gigantic project of water engineering that will drive a million people from their homes is the building of the Three Gorges project on the Yangtze River in China.

Western consortia of companies have waited for years for the contracts worth 4 billion US dollars, for the dam intended to control the floods that periodically destroy the lives and livelihoods of vast numbers of people in the fertile basin beyond the Gorges. It will create a 500-kilometre-long reservoir deep in western China. A second, powerful argument for the proposal is the prospect of hydro-electric power on an immense scale. As Pearce puts it:

> *China, like India, has projections for its future energy demands that make environmentalists blanch. If it concentrated on exploiting coal reserves, China could on its own double the worldwide greenhouse effect within 50 years. From this perspective, the attractions of pollution-free hydro-electric power are obvious.*[15]

He believes that if the Chinese government is anxious to spend all that money, half should be invested in using power more efficiently and the other half on 'furthering its already remarkable programme of constructing small run-of-river hydro-electric plants'.[16] Other Western observers stress the vulnerability of the dam to accident, unforeseen climatic events, earthquakes and landslides, sabotage and war, as well as the build-up of silt experienced with the Aswan Dam. They also claim that China's Western consultants have deliberately understated the human consequences of the project.[17]

In the Indian subcontinent, South-East Asia, Africa, the Middle East and Latin America, the super-power that local communities fear most is not the ruthlessness of oriental despotism but the combination of central governments and international funding agencies like the World Bank. This is an agency of the United Nations, borrowing money and lending it at commercial rates.

This gives it an automatic bias in favour of the values of market economics, promoting, in the case of large-scale water engineering, projects which promise increased production for export markets rather than for the local subsistence economy. Far from being an organization giving priority to the human needs of the people whose lives it disrupted, it has followed the path of the British cotton manufacturers' support for the first Aswan dam or the Volta River project in Ghana resulting in the American Kaiser Corporation winning electric power for 5 per cent of the average world price for aluminium production at a huge but hidden cost to the rural poor.[18] Many observers of the issues facing the countries of what became known as the 'developing' world hoped that the aims of United Nations intervention would be more closely linked to the lives and hopes of existing populations.

The World Bank and the International Monetary Fund were founded at the Bretton Woods conference in 1944, pledged to avoid the mass unemployment of prewar years and to promote international trade. Commenting on the fiftieth anniversary of the founding conference, the unofficial body Oxfam, which promotes relief and development programmes in over 70 countries, observes that

The Bretton Woods system was created, with the experience of Germany in the inter-war period fresh in mind, to protect employment and regulate markets without recourse to extreme deflationary policies. Yet World Bank and IMF structural adjustment programmes, developed in response to the Third World debt crisis of the 1980s, have concentrated on achieving low inflation and deregulating markets to the exclusion of other considerations. The resulting deflationary pressures have undermined prospects for economic recovery. They have also contributed to high levels of unemployment and the erosion of social-welfare provision for the poor. Meanwhile, market deregulation has brought few benefits for those excluded from markets by virtue of their poverty and lack of productive resources.[19]

In the field of water management, the World Bank, over its fifty-year history has favoured lending for vast, spectacular projects, with little regard for the local human and ecological consequences. Oxfam complains that 'Negotiations with governments are held in secret, and non-governmental organizations and citizens' groups are denied access to information.'[20] The same complaint is made by other bodies in trying to assess the World Bank's slowly developed techniques for assessing the impact of the projects it funds. They complain that only five full-time professionals in the Bank's Office of Environmental Affairs, among 5,250 employees, are concerned with its environmental review procedures:

> *On these five people rests the impossible task of monitoring hundreds of ongoing projects and examining the approximately 300 new projects that are approved each year . . . As told by the Natural Resource Defense Council's Bruce Rich to a US congressional hearing, 'the Bank's sole mandatory environmental procedures call for the review of projects long after they have been chosen and designed, and just before the initiation of loan negotiations.*[21]

Every few years, faced by the manifest failure of projects it has financed to show either an economic return or to produce benefits for the populations affected, the World Bank conducts a reassessment of its activities, fully admitting the validity of the complaints of its critics and promising to do better in future. Thus in 1993 it reported that 'In many of the projects surveyed, consultation with affected populations and local non-governmental organisations has been limited at best. Where it did occur, it helped improve project design and correct local misconceptions and their impact.'[22]

And in the following year, celebrating the fiftieth anniversary of the Bretton Woods conference, the Bank committed itself to 'increased awareness of and sensitivity to the social and ecological dimensions of development', and to 'enhance the participation of the poor in the design and implementation of Bank-financed projects and programmes'.[23]

The Lure of the Dam

The irony of these resolutions is that they coincide with intense local opposition to vast adventures in dam-building and water engineering proposed by central governments with the promise of World Bank finance.

A turning-point came with the worldwide publicity given to local opposition to just two of the Bank-funded projects on the Indian subcontinent. The Narmada valley project in western India is, in Fred Pearce's words, many times larger than its distant ancestor, the Tennessee Valley project. The Sardar Sarovar Dam and a further projected Narmada Sagar Dam were intended to feed the largest lined dam in the world, which

> would take water from Sardar Sarovar across the arid state of Gujarat almost to the border with Rajasthan. Gujarat's leaders claimed that poor drought-prone villagers, and especially tribal groups, would receive its water, but in fact the canal by-passes most of those villagers. Instead, it goes to districts containing rich landlords growing water-hungry cash crops, and the industrialists of Ahmedabad. [24]

Medha Patkar, the woman who began a campaign simply to inform villagers of their rights, found herself leading a movement for reversing the attitude of the World Bank. The Bank commissioned an advisory report which condemned the project, and its American, Japanese and German directors voted against continuing support. But they were a minority. An international pressure group accused the Bank of failing 'to acknowledge the widespread opposition and associated human rights abuses', [25] and in April 1993 the World Bank resolved not to put further funding into the project. The only foreign funder still involved was the British Overseas Development Agency. The Gujurat and central governments declared that they would 'go it alone' with the completion of the vast undertaking. [26]

The big turn-around in the World Bank policy towards water engineering was the proposal for the hydro-electric project known as Arun III in Nepal. This project is totally different in scale from the Narmada dam. It would affect only 155 families, rather than the 155,000 people touched by the Indian project. It would cost one billion dollars rather than 11 billion. Supporters claim that it 'would provide more than 400 megawatts for a country in desperate need of electricity. So far, Nepal has developed only 241 megawatts of hydro-electric power, accessible to barely 10 per cent of the population'; while objectors see it as 'unnecessarily large and expensive, diverting resources away from social programmes like health and education'. It is in a remote valley, 200 kilometres east of Katmandu 'inhabited by endangered species such as the Asiatic black bear, the clouded leopard and the Annamese macaque'.[27]

Villagers too are an endangered species. For rural Nepal has a fuel crisis. 'Journeys to gather firewood and fodder in the Himalayan foothills now take a whole day, compared to an hour or two a generation ago', while deforestation is held responsible for the area of flooding from Nepalese rivers increasing four times in the past 25 years.[28] But the local people would not benefit, since the power is destined for the urban elite, among that small proportion of the population who already have access to electricity.

Finally, in August 1995, the World Bank cancelled its plan to fund the Arun project. Welcoming the announcement, British group Intermediate Technology stressed that Nepal's energy problem still exists and drew attention to its proposal to the Bank in 1993 for a feasibility study of possible small-to-medium-scale hydro sites, exploiting Nepali skills in developing sustainable, economically stable power sources for the future.

Large-scale water engineering, apart from fomenting international disputes, has ceased to be a Cold War weapon, but has become an economic bargaining counter. The example that became notorious, simply because it was exposed, was the British government's funding of the Pergau Dam in Malaysia. In 1994, on the day after the High Court in London had ruled that the British government's decision to spend £234 million of aid money on the Pergau dam project in Malaysia was unlawful, the press explained the background:

> *Baroness Thatcher, then prime minister, promised to provide financial help for the dam in 1989 while negotiating a £1.3 billion arms deal for Tornado jet fighters, artillery, radar, submarines and missiles. The following year the Overseas Development Administration, which is responsible for Britain's overseas aid programme, concluded that the project was poor value, costing £56 million more than was needed.*

Malaysia subsequently ordered 28 British Aerospace Hawks worth £403 million. Then, in March 1991, the £400 million contract to build the dam was won by an all-British joint venture. The Foreign Secretary, Douglas Hurd, authorised the first payment of aid in July 1991 despite more advice from the permanent secretary at the ODA that the dam was not a sound development project.[29]

No ministers felt obliged to resign because the government has misused its aid budget, and payments to the Malaysian government will continue. The case had been brought to court by the World Development Movement, an unofficial body established because of new laws prohibiting aid charities from involvement in political lobbying. In effect, it represents the interests of charitable aid agencies like Oxfam, Christian Aid and Action Aid, whose existence indicates the gulf between the two levels of water management, governmental and local.

Modern large-scale water management began in countries with ample rain and a temperate climate, and where regard had to be taken of the interests of existing populations. It has been exported all over the world to countries with a host of interrelated factors, summed up by Joan Davidson as deforestation, intensive land use, centralized planning and inequitable land distribution,[30] which ensure that local human populations, quite apart from the local ecology, are not the beneficiaries. Technological imperialism has replaced the empire-building of the past.

When European powers conquered the cities of Africa and Asia, they built themselves new settlements, laid out with military precision at a distance from the old city, taking care to provide a clean water supply and safe disposal of sewage, training a new servant class, and trying to insulate themselves from the endemic water-borne diseases that afflicted the native population. But, as in Victorian England, the rich could not isolate themselves from the daily hazards of the lives of the poor, and their children were in daily contact with a much-loved local *ayah*. Something must be done about the old city.

In India, engineers employed by the Indian Civil Service devised plans for filling in the traditional local water tanks with the aim of combating malaria, even though they absorbed excess water in the monsoon season. They planned piped water supplies from new distant reservoirs and dams, and sought to replace the daily removal of faeces in the sweeper's cart by a water-borne sewage system, which, even if a water supply were available, 'was estimated to cost twice the value of all the house-building in the town'.[31] This was to be accompanied by wholesale demolition of slum areas. Needless to say, such proposals would penalize the poor, would destroy the fabric of the city and would be grotesquely expensive.

During the First World War the Scottish biologist Patrick Geddes, whose concept of regional planning was described in Chapter 3, finding no work at home, was commissioned by a variety of municipal and local maharajahs to prepare planning recommendations for a series of towns and cities in what are are now the republics of India, Pakistan and Bangladesh. These reports had a very small circulation at the time, some exist only in typewritten form, but the published accounts of them illustrate an approach to the water problems of poor communities which is based on community action rather than on indebtedness to international contractors.

In one of his series of Madras reports Geddes urged that revival, rather than neglect or filling, should be the policy towards the urban water tanks:

This filling up of tanks, created by ancient foresight, labour and sacrifice, seems too often lightly suggested. That the edge be regularised, and the slope of the bottom also, so as to avoid the irregular development of stagnant and mosquito-breeding pools and to admit of the steady retreat of the water towards a central and deeper portion – that is surely the more practical policy, and an enormously cheaper one. By thus arranging a pool of retreat, loss by evaporation would be diminished, as well as mosquitoes abated. And if annual drying up prevents keeping up the supply of fish to keep down larvae we may efficiently replace fish by ducks, whose incessant searchings in mud and water, on bottom and on the surface, are so peculiarly thorough and efficacious; while even if complete drought supervenes, the ducks survive. Is it not part of that curious apathy to minor agricultural interests which is so common amongst the educated classes, Indian not less than European, that such simple aid against one of the gravest scourges should not be provided and maintained?[32]

Geddes brought the standpoint of a biologist, rather than that of a civil engineer, to the overwhelming problems of Indian cities. He was in touch with both Tagore and Gandhi, and like them, saw as a major task the upgrading of the status of the 'sweeper', whose cart should 'pass frankly along the main thoroughfare. When conservancy lanes are provided for the cartage of ordure, the system falls steadily and surely to the level of its purpose. When the sweepers pass along the main streets, both their methods and their standards rise.'[33] He sought not sewage farms, but local composting and fertilized gardens, converting what would have been 'a fetid and poisonous nuisance into a scene of order and beauty'.[34]

Geddes looked at water issues from the standpoint of the poorest citizens and passed his observations to local authorities who would have felt far more at home with recommendations for vast engineering works which they could not afford. He wanted to involve local populations in water festivals which 'improved and heartened up' the neighbourhood, instead of leaving it 'annoyed, alarmed, embittered or depressed', rather than in impositions from above which kept down 'their health-consciousness and health-conscience to the level of the rectum'.[35]

This approach to water as a community resource, based on local water-gathering, and local conversion of wastes, so that every river valley could use its share of resources, had to be rediscovered many decades later. But not before remote rulers, spending borrowed money on projects from world-ranging engineering firms, had guaranteed that community co-operation would be replaced by international disputes over water management.

Notes

1. Fred Pearce, *The Dammed: Rivers, Dams and the Coming World Water Crisis* (London: The Bodley Head, 1992), p. 345.

2. Percival Goodman, *The Double E* (New York: Anchor Books, 1977), pp. 132–3.

3. Welsh Nationalist Party, *TVA for Wales* (Caernarfon: Welsh Nationalist Party, 1942).

4. Thomas Johnston, *Memories* (London: Collins, 1952), p. 130.

5. Peter Hennessey, *Never Again, Britain 1945–1951* (London: Johnathan Cape, 1992), p. 206.

6. Johnston, *op. cit*, p. 178.

7. W. H. Murray, *The Companion Guide to the West Highlands of Scotland* (London: Collins, 1968) p. 132.

8. Harold Barclay, 'Anthropology and anarchism', *The Raven*, Vol. 5, No. 2 (April–June 1992), pp. 170–1.

9. Pearce, *op. cit*, p. 78.

10. *Ibid.*, p. 81.

11. *Ibid.*, p. 79.

12. *Ibid.*, p. 285. See also M. Lavergne, 'The seven deadly sins of Egypt's Aswan High Dam', in E. Goldsmith and N. Hildyard (eds), *The Social and Environmental Impact of Large Dams*, vol. 2 (San Francisco: Sierra Club, 1984).

13. Robin Clarke, *Water: The International Crisis* (London: Earthscan Publications, 1991), pp. 59–62.

14. Pearce, *op. cit*, p. 3.

15. *Ibid.*, p. 240.

16. *Ibid.*, p. 241.

17. Gráinne Ryder (ed.), *The Three Gorges: What Dam-Builders Don't Want You to Know* (Toronto: Probe International; London: Earthscan Publications, 1990). See also Audrey B. Topping, 'Three Gorges gamble', *Foreign Affairs* (September–October, 1995).

18. David Hart, *The Volta River Project: A Case Study in Politics and Technology* (Edinburgh University Press, 1980).

19. Oxfam Policy Department, *A Case for Reform: Fifty Years of the IMF and World Bank* (Oxford: Oxfam Publications, 1995), pp. 1–2.

20. *Ibid.*, p. 3.

21. Patricia Adams and Lawrence Solomon, *In the Name of Progress: The Underside of Foreign Aid* (London: Earthscan Publications, 1985. Citing the US Congressional Hearings of the Subcommittee on International Development Institutions and Finance, June 1983.)

22. World Bank, *The World Bank and the Environment 1993* (Washington: World Bank, 1993).

23. World Bank, *Learning from the Past, Embracing the Future* (Washington: World Bank, 1994).

24. Pearce, *op. cit*, p. 157.

25. *New Scientist*, 31 October 1992.

26. *Guardian*, 16 April 1993.

27. Mark Tran, 'World Bank sees Nepal project as test of credibility', *Guardian*, 7 November 1994.

28. Adams and Solomon, *op. cit*, p. 127.

29. 'High Court rules money for dam project was unlawful', *Guardian*, 11 November 1994.

30. Joan Davidson and Dorothy Myers, *No time to Waste: Poverty and the Global Environment* (Oxford: Oxfam, 1992).

31. H. V. Lanchester, Preface to Jaqueline Tyrwhitt (ed.), *Patrick Geddes in India* (London: Lund Humphries, 1947), p. 19.

32. Cited in Helen Meller, *Patrick Geddes, Social Evolutionist and City Planner* (London: Routledge, 1990), p. 240.

33. *Ibid.*, p. 254.

34. *Ibid.*, p. 255.

35. *Ibid.*, p. 239.

CHAPTER 5

Fighting over Water

> *Engineers involved in water planning will often avoid situations where public debate and participation is necessary, preferring to see planning as a design process for which mathematical formulae provide acceptable answers. Consultation therefore often follows after the scheme is 'perfected', according to traditional engineering designs, instead of embracing political debate about scheme aims or policy goals. Any 'subjective' data are eliminated leaving the planning process comfortably neutral but, unfortunately, incomplete.*
>
> **Dennis Parker and Edmund Penning-Rowsell,**
> **Water Planning in Britain.**[1]

People in pubs and clubs enjoy quoting Mark Twain's remark that whiskey is for drinking but water is for fighting over. This invariably provokes someone to tell the tale that is part of the folklore of the water supply industry: that of the vendetta between the water company and the brewery. The beer-makers were big industrial customers of the company which processed water from its reservoirs fed by the river and its upstream boreholes. But they discovered that modern drilling techniques and sophisticated geological surveys enabled them to bore deep under their site and win pure water of their own. The aggrieved supplier retaliated by boring deeper and draining the brewery source. The brewers responded with an even more ingenious offensive in the water war, followed by another counter-attck until, eventually, after huge expenditure by both parties, an agreement was reached.

The reason why people like to tell this tale is obvious. It provides a microcosm of far more dangerous disputes based on the exploitation of water.

Yet Jean Robert argues that the sharing of water resources by mutual agreement between communities is a fact of history.[2] He is a Swiss-born architect who worked at his profession for several years until he paused to reflect that his work was 'dedicated to a large extent to the construction of banks'. As a result of contact with Ivan Illich and John Turner, he moved to Cuernavaca in Mexico, where for twenty years his main practical interest has been in promoting, at grassroots level, safe non-waterborne systems of sewage disposal. His theoretical concern has been to discover the principles that should govern our use of water, balancing conservation with the need to guarantee access for the poor and vulnerable.

For him, as for the regionalist prophets discussed in Chapter 3, the valley of a river and its tributaries is the 'natural' unit for sharing out water. This view was upheld by the International Conference on Water and the Environment in 1992, which concluded that 'the most appropriate entity [for water politics to be effective] is the river basin, including surface and groundwater'.[3]

From this proposition Robert concludes that those courageous village communities objecting to the vast water engineering projects all over the poor world are evoking that natural right which he describes as the first golden rule of all water policy;

> *No new waterworks — or transportation, energy or other 'developments' for that matter — should ever be proposed if the affected community's right to say no to them has not been clearly recognised and if the non-realisation of the project is not publicly debated as a concrete option.*[4]

He expresses it this way because he is aware that the token consultation that is sometimes insisted on by funding bodies like the World Bank is a meaningless procedure if it is taken for granted that, having listened politely to the objectors, the developing agency is determined to go ahead anyway.

If this first golden rule were taken seriously, Robert claims, the risks of war generated by disputes over water would not occur. He is aware that 40 per cent of the world's population depends on water from a neighbouring country, and

> *more than 200 large rivers are shared by two or more countries. One country's hydro-electric, irrigation and water supply projects may cut*

off a neighbour's water supply. But has water sharing been a major cause of war in past times? Upon careful investigation, the opposite might appear to be true.[5]

For he reaches the memorable conclusion that 'Throughout history, water has been a motor of peace rather than of war. Since time immemorial, people riparian of the same watercourse have learned to make peace by concluding agreements about the use of their shared water.'[6]

In the modern world, the existence of vast cities, the spread of irrigated agriculture, and the demand for hydro-electric power, have complicated the principle of basin-consistency. We can see this in a densely populated country like Britain with a long tradition of large-scale water management. Just as Liverpool draws water from Lake Vyrnwy in Wales, so Birmingham's water supply comes by gravity from five reservoirs draining 70 square miles (182 square kilometres) of rainy mountains above the Elan valley in Powys. News of massive diversions upstream, however 'basin-consistent' they were, would bring consternation to Birmingham's water planners. We might be confident that discussion, reference to existing agreements and common sense would resolve the issue, settled amicably to the satisfaction of all parties.

But in many parts of the world large-scale water manipulation worsens both internal and international tensions. Stephan Libiszewski, who studies the resolution of environmental conflicts at the Swiss Federal Institute of Technology in Zurich, explains that international law

> *does not provide adequate means to regulate the competition between riparian states. Upstream states can refer to the doctrine of absolute national sovereignty, whereby a state has the exclusive right to use and dispose of the natural resources within its territory. Downstream states tend to emphasise another principle: the doctrine of absolute national integrity according to which lower riparians are entitled to unaltered water volume and quality. Given these contrasting doctrines and modern abilities to dam and divert rivers, international conflicts over the sharing of transboundary water resources are almost inevitable.*[7]

The question that arises for him, is the same one that occurs to Jean Robert: why do states in arid regions fail to co-operate in water management and development when co-operation would appear to be in their mutual interest? He cites the various drafts known as the Helsinki Rules, which try to establish the criterion of 'equity', but concludes that water disputes will not be solved 'until the overwhelming political and territorial conflicts have been settled'.[8] Large-scale engineering projects and their funding depend on engineering concepts of 'efficiency' and of the anticipated 'yield'. They inevitably minimize what are seen as the losses in efficiency that result from exploiting less than the maximum capacity and of respecting the needs of other water-users.

This can be seen all over the Middle East. Countries downstream in the Euphrates–Tigris basin, like Syria and Iraq, feel threatened by the impact of the vast Ataturk Dam which intermittently cuts off the flow of the Euphrates, while 'One of the reasons why the Turks are attacking the Kurdish rebels is that water resources in the Kurdish area are important for Turkey.'[9] One aspect of the Egyptian government's Aswan High Dam is that the Nile basin includes eight other African countries, especially its immediate upstream neighbours. Robin Clarke noted the significance of the fact that Israel's water-transfer agreement with Egypt 'offered Israel 400 million cubic metres per year of fresh water in exchange for a Palestinian solution'.[10] On that occasion President Sadat remarked that 'The only issue that could lead Egypt to war is water,' but his warning was addressed to Ethiopia.[11]

Some water-watchers feel that they have waited a lifetime for the concept of equity and of water as a common good to re-emerge in the Middle East. Forty years ago I attended an international conference in London on Regional Planning and Development. It was privately organized by members of a defunct organization, the Association for Planning and Regional Reconstruction, and, for unfathomable reasons, was smeared by the British government as a 'Communist Front'. The assistant general manager of the TVA read his paper and withdrew on instructions from the US Embassy. The Colonial Office speaker's paper on the Volta River Project, in what was then the Gold Coast and is now Ghana, was read for him as he had been told to keep away. I reported the discussion of his paper, and am struck by the later relevance of the comments made at the time. The criticism made, when the dam was still a dream of the British and Canadian aluminium companies and the British and Gold Coast governments, have been amply justified by subsequent history.

> Mr A. L. Bryden, a lawyer, spoke of the difference between our own law of land tenure and African tribal law which did not recognise the existence of land as a commodity but recognised the right to the use of the earth and that which it produces; it belonged to the community. Mr de Schlippe, an authority on tropical agriculture, declared that no attempt has been made to solve the problems of the remaining rural population, nor was there any effort to make way for the growth of 'a natural spontaneous organisation to sort out these problems'. Dr Otto Koenigsberger agreed with him on the new social problems created by 'an island of very high technological development'. Mr T. Baloch, the economist, said that 'It is appalling how the financial and economic aspects are being treated just as a matter of high-powered bookkeeping.' And why, he asked, was the control of the proposed aluminium firm which so vitally affected the welfare of the people, to be left in the hands of the aluminium companies who were not providing a large share of the capital and might use the Volta simply as a buffer plant, to be switched on or off with every change in market conditions?[12]

Long after Ghanaian independence, the hidden costs of the dam were discovered. As Fred Pearce puts it, the waters of the Volta had been sold very cheaply, and 'they had spread disease across the countryside, brought chaos to stretches of the coastline and drowned the homes and farms of 80,000 people'.[13]

On the walls at that same conference in 1955, the Tennessee Valley Authority displayed the plan it had prepared at the request of the United Nations, called *The Unified Development of the Water Resources of the Jordan Valley Region*. After the partition of Palestine and the subsequent fighting, the UN Palestine Conciliation Commission sought this 1953 plan for the area. The TVA team did not visit Palestine at all, but relied on engineers' reports.

> In March 1954 the Arab Plan for *The Development of the Water Resources of the Jordan Valley* was issued. It was the first all-Arab scheme for regional water development and it recognised Israel's right,

> *for the first time, to a share of the Jordan water. Meanwhile Israel submitted the so-called Cotton Plan made by an American, Mr J. S. Cotton; this plan also provided for sharing the Jordan water. By the spring of 1955 the problem of competitive claims on Jordan water had been simplified by a detailed engineering survey made jointly by M. Baker of Rochester and the Haza Company of Chicago with approval of the Jordan government and at the request of the US Foreign Operations in Jordan. The upshot of this study showed that considerably less water would be required per acre than had previously been supposed . . . Differences between the Arabs and Israelis were considerably narrowed. Both sides were agreed on the necessity for a joint scheme and on the utilisation of Lake Tiberias as a main reservoir, and to some form of international supervision.*[14]

The most disheartening thing about this account of a plethora of engineering plans is the forty years of conflict since it was written. Attempts to divert the sources of the Jordan in South Lebanon and the Golan Heights provoked the Israeli–Arab war of 1967, and although a working peace has been achieved with neighbouring states in the Jordan Valley, there is no agreement over fair access to water for the Palestinians on the West Bank who, since the occupation began in 1967, have been barred from digging new wells or renovating old ones. In 1995, commenting on current water negotiations, *The Economist* reported that

> *Each year Israel pumps 600m cubic metres of water (over 30 per cent of its supply) from aquifers that lie, partly or wholly, under the West Bank. Of this, 115m cubic metres is allocated to the West Bank's 1.4m Palestinians and 30m to the 130,000 Jewish settlers there. The rest goes to Israel, servicing Jerusalem and greater Tel Aviv.*[15]

In the early decades of this century, long before the foundations of a state of Israel, pioneer settlers won the affection of Arab neighbours by sharing what were then new techniques of water-gathering with them. It was as long ago as 1920 that the philosopher Martin Buber warned that if the incomers did not live *with* the existing population as well as *next* to them, they would find themselves living in enmity towards them.[16]

Conflict over water supplies may be only one of the issues that lead rulers to draw their subjects into war, but Ismail Seregeldin, vice-president of the World Bank, declared that 'The wars of the next century will be over water.' He reminded us that 'By the year 2025 the amount of water available to each person in the Middle East and North Africa will have dropped by 80 per cent in a single lifetime.'[17]

It is not simply a matter of competition for supplies of a limited resource, for the fact is that modern wars depend upon the destruction of the civilian population's means of life and livelihood. Joan Davidson and Dorothy Myers provided a shocking account of the environmental consequences of the 1991 Gulf War. Not only did 'Black rain' fall for months on Kuwait, Iraq and Iran following the firing of oil wells by the retreating Iraqi army, but

> *The social and environmental effects of the Gulf War on Iraq's agriculture and water supplies have been devastating. The breakdown of electric power supplies following Allied bombing has created a cycle of contamination. More than 90 per cent of the sewage treatment plants in the country are out of action, which means massive amounts of untreated domestic and industrial sewage are pumped into the rivers. A Harvard Study Team which visited Iraq in 1991 reported that 'In all of the seven southern governorates surveyed, the onset of insanitary conditions and the increase in water-borne diseases followed the loss of electric power in the first days of the war.' Agricultural production has been sharply reduced as a result of the breakdown of the electrically-powered network of irrigation pumps . . . Before the war, Iraq produced 30 per cent of its food needs. Today it produces 10–15 per cent.*[18]

The powerful make war and the powerless suffer the consequences. But the Gulf War was in no way related to rivalry over water. Kuwait is a desert land, heavily dependent on the most costly of all means of water extraction, the desalination of seawater. In many parts of the world, far from the Middle East, international hostilities are exacerbated by disputes over water resources. Governments, in stirring up support for external belligerency, have only to cry, 'Look! They are stealing *your* water!' for populations to press for military action.

The rich United States and poor Mexico have, throughout this century, maintained a continuous dispute over the water flowing in the basins of the Rio Grande and the Colorado River. That it has not resulted in warfare is simply because of the vast disparity in military might between the two countries. On both sides of the border irrigated farming has been used increasingly, drawing on boreholes to underground aquifers. This has resulted not only in a significant lowering of the water table, but in continually increasing salinity in the water flowing through the rivers on the American side into Mexico. Conflict is increased by the fact that in the United States the Texan farmers have absolute rights over the water flowing through their land, while in Mexico it is seen as a national resource, controlled by rules made by the federal government in Mexico City.

The rivers of South America begin in the Andes and flow down to broad valleys and coastal plains before discharging into the Atlantic or the Pacific oceans. They all pass through several nation-states, and engineering projects in one country upstream can have devastating effects for downstream farmers in another country threatened by flooding. On the Pacific side Bolivia and Chile have been in a continuous dispute over the use of the River Lauca, as a result of water diversions for hydro-electricity and irrigation. On the Atlantic side the immensely larger river system of the Rio de la Plata and the River Paraná flows through five countries, and while the basin occupies much more of the territory of neighbouring lands, Brazil, where the rivers rise, controls their use upstream and has embarked upon several huge, internationally financed, engineering and dam-building projects. Successive Brazilian governments have frequently failed to inform their downstream counterparts of their intentions.

Current observers of these adventures are scathing. Robin Clarke notes how 'Relations between Brazil and Argentina soured over the affair, and opponents of the dam, including Brazilians, claimed that the development bore less relation to the country's energy needs than to Brazil's "militant posturing" towards neighbouring countries.'[19] Argentina followed Brazil's example and set about the Yacyreta hydroplant stretching across the floodplain of the River Paraná, which will extend for 70 kilometres from Paraguay to Argentina. Fred Pearce comments that

No icon this, it has been plagued with design and construction problems and was described by the Argentine's President Menem as a 'monument to corruption' in the former dictatorship. Costs of a project that will displace more than 100,000 people have risen from $1.5 billion to $6.5 billion, and the final bill may be $12 billion.[20]

The appalling consequences of large-scale governmental visions of water control can be seen in the ever-present threat of water wars in South Asia. They add fuel to the flames of resentment between India, Bangladesh, Nepal and Bhutan, through which the Brahmaputra and Ganges flow. For, apart from the effect of big dams, already threatened by the erosion of topsoil, large-scale deforestation upstream brings continually more widespread flood disasters downstream, where the flood-plain is now faced by the new threat.

On the other side of the subcontinent the waters of the Beas-Sutlej and Ravi rivers have been a continual source of conflict. Robin Clarke provides a succinct account of the immensely complex confrontation resulting from big governmental plans for this vast river basin:

These rivers rise in the Himalayas, and flow through the predominantly Sikh state of Punjab in India, into Moslem Pakistan. To the south of Punjab are the predominantly Hindu Indian states of Rajasthan and Haryana which do not have their own river system. In 1948 India diverted the course of the Beas-Sutlej and Ravi rivers so that they no longer flowed into Pakistan. This happened in the spring growing season and threatened Pakistan's agricultural output for the whole year. The dispute was settled only through the intervention of the World Bank. Punjab water was an important contributing factor in conflicts such as the 1965 Indo-Pakistan war, and the Indian storming of the Sikh Golden Temple in Amritsar in 1984. The Punjab presently gets about 40 per cent of India's agreed share of the Beas-Sutlej-Ravi River water, and the Sikhs' main political group is asking for a greater amount of the river water to be made available to the Punjab for irrigation. The water issue has become inextricably linked with the wider religious and political demands that led to the violent confrontation at the temple in Amritsar.[21]

Mr Clarke's description of the situation will probably satisfy adherents of none of the factions involved, and like all the other accounts of current international disputes over water, it appears to contradict Jean Robert's claim that all through history people along the route of the same watercourse have learned to make peace through their common use of their shared river. But this is partly because of the intervention of the modern concept of the sovereign nation-state, flexing its muscles in opposition to another governmental machine across the border. It results even more from the sheer size of twentieth-century engineering products. The exercise of restraint, and of exploitation of less than the maximum flow in the interests of other users of the same river, is foreign to the concept of maximum use of any source of water. It could make the difference between a project which is considered financially viable, and one which is not a commercial proposition. So technical solutions take precedence over human solutions.

People are ill-served by governments, and governments are ill-served by the consultants and contractors they commission. There is an alternative approach to water needs which begins at the opposite end of the water spectrum, with the local community and its needs, and with direct control of the use and sharing of a precious resource. This inherited tradition was described in Chapter 2. Its application to another of the great river systems of South Asia is described by Bharat Dogra in a report dedicated 'to the anonymous struggles and their unnamed heroes who struggle at the village level against floods and their causes in South Asia'.[22] His little book on *Floods in South Asia* was written in the aftermath of the floods of 1992, with a loss of about 10,000 lives, and describes the Ganga-Brahmaputra-Barak river system, where nearly half a billion people live (almost one-tenth of the world's population.)

This vast series of river basins is spread across five nation-states – India, Bangladesh, Nepal, Bhutan and Tibet (China) – and Bharat Dogra explains that

> *This region known once for the fertility of its land, abundance of its fisheries, skills of crafts persons, has been forced into large scale poverty due to centuries of colonial exploitation made worse by distorted development of post-colonial days. The result is that the*

region is today known more for its poverty, hunger and disasters than anything else.[23]

He reports on deforestation and reforestation in the upland regions, on soil erosion and on large-scale projects for flood protection. The last people to be consulted or trusted were the local populations at any stage in the river systems, and the concept of the nation-state complicates the possibility of water sharing:

> *In fact the complex evidence that has been accumulating about the longer term impact of large dams, barrages and other big engineering structures makes it clear that any group which manages to get more water, or any other such obvious advantage, will not necessarily be the main gainer. For example, let's assume that a reservoir is to be created in country 'A' but most of the land to be irrigated lies downstream in country 'B'. So people of country A are made to sacrifice their land to irrigate the land of country B. but if the dam is an unsafe one, after some time it collapses and kills thousands of people in country B. This can happen if country B ignores the more basic question of dam safety and instead only tries to get a better deal vis-à-vis country A. While so far we neighbours have been suspicious of each other, we've had no hesitation in inviting any number of foreign consultants and companies and giving them millions of dollars for jobs which we could have done ourselves by helping each other.*[24]

This is why, reflecting the views of the South Asian People's Environment Network, he urges 'small projects that show a trust of each other', rather than large-scale projects bringing huge international debts owed by poor nations to rich nations, and which involve the displacement of millions of people in fields and slums.[25] If we could trust each other, we would have no incentive to fight over water.

Notes

1. Dennis Parker and Edmund Penning-Rowsell, *Water Planning in Britain* (London: Allen & Unwin, 1980) p. 245.

2. Jean Robert, *Water is a Commons* (Mexico D. F.: Habitat International Coalition, 1994).

3. International Conference on Water and the Environment, *The Dublin Statement* (Dublin, 31 January 1992).

4. Robert, *op. cit*, p. 90.

5. *Ibid.*, p. 30.

6. *Ibid.*, p. 31.

7. Stephan Libiszewski, 'Water, water, everywhere . . .', *The Ecologist*, Vol. 24, No. 5 (September–October, 1994), pp. 196–7.

8. *Ibid.*, He was discussing Nurit Kliot, *Water Resources and Conflict in the Middle East* (London: Routledge, 1994); John Bulloch and Adel Darwish, *Water Wars: Coming Conflicts in the Middle East* (London: Gollancz, 1993); and Miriam R. Lowi, *Water and Power: The Politics of a Scarce Resource in the Jordan River Basin* (Cambridge University Press, 1993).

9. Torvild Aakvaag, 'Can the needs of society and the environment be reconciled?', *RSA Journal*, Vol. CXLIII, No. 5464, November 1995, p. 38.

10. Robin Clarke, *Water: The International Crisis* (London: Earthscan Publications, 1991), p. 101.

11. Robert, *op. cit.*, p. 30–1.

12. Colin Ward, 'The Conference on Regional Planning', *Freedom*, 8 October 1955, p. 5.

13. Fred Pearce, *The Dammed: Rivers, Dams and the Coming World Water Crisis* (London: The Bodley Head, 1992), p.126.

14. W. E. R. Gurney, 'JVA – a Jordan Valley Authority?' *Doorway to the 20th Century*, Vol. 2, No. 3 (March 1956), pp. 21–32.

15. 'Whose water?', *The Economist*, 5 August 1995, pp. 52–3.

16. Martin Buber, *Israel and Palestine* (London: East & West Library, 1951).

17. John Vidal, 'The water bomb', *Guardian*, 8 August 1995.

18. Joan Davidson and Dorothy Myers, *No Time to Waste: Poverty and the Global Environment* (Oxford: Oxfam, 1992), p. 133.

19. Clarke, *op. cit*, p. 98.

20. Pearce, *op. cit*, p. 138.

21. Clarke, *op. cit*, pp. 94–5.

22. Bharat Dogra, *Floods in South Asia: A Report on Ganga-Brahmaputra Region* (New Delhi and Dhaka: South Asian People's Environmental Network, 1993).

23. *Ibid.*, p. 3.

24. *Ibid.*, p. 42.

25. *Ibid.*, p. 43.

CHAPTER 6

Small and Local

Poor people have relatively simple needs, and it is primarily with regard to their basic requirements and activities that they want assistance. If they were not capable of self-help and self-reliance, they would not survive today . . . If we clearly understand that one of the basic needs of many developing countries is water, and that millions of villagers would benefit enormously from the availability of systematic knowledge on low-cost self-help methods of water-storage, protection, transport, and so on, there is no doubt that we have the ability and resources to assemble, organise and communicate the required information.

E. F. Schumacher, *Small is Beautiful*[1]

When Fritz Schumacher and George McRobie started the Intermediate Technology Development Group (ITDG) in 1965, they were both economists working for the National Coal Board in Britain, and were frequently asked to advise on the economic problems of countries then seen as the Third World. In those days it was thought that access to high technology held the secret of success for all nations, great and small. They grasped the fact that then, as now, the dilemma of the poor regions was their dependence for imports on the export of cash crops and raw materials. If they tried to diversify, they found that there was always some other country which could produce more cheaply. If they sought to build up their own modest production units to supply local needs, on the Gandhian pattern, they found that another country could mass-produce the same goods and deliver more cheaply too, while the industrial plant they would like to buy was geared to the advanced technology of the rich countries.

Would-be purchasers in the poor world would say to McRobie and Schumacher, 'There used to be a piece of equipment you could buy for £20 to do a particular task. Now it is fully automated and costs £2,000 and we can't afford to buy it.' So they set up the ITDG to locate or design machines and tools and techniques that would help villagers and poor city-dwellers with a superfluity of labour and a shortage of capital. For they say that, in countries plunged into debt by huge loans for sophisticated enterprise, the poor and their urgent daily needs were forgotten or ignored. And they accumulated a series of little local maxims:

'If you want to go places, start from where you are.'
'If you are poor, start with something cheap.'
'If you are uneducated, start with something simple.'
'If you live in a poor environment, and poverty makes markets small, start with something small.'
'If you are unemployed, start using your own labour power, because any productive use of it is better than letting it lie idle.' [2]

This approach sounds like the folk wisdom that most people absorb from their grandparents, but is a world away from the habit of thinking big which characterizes governments everywhere. And in no field of life is the contrast more startling than that of water, whether as a necessity in food production or as the basis of personal and domestic life. In the modern world it is possible for people to have access to the internal combustion engine, radio and television, but not to a safe water supply. Bangalore, for example, the home of India's computer software industry (where software engineers cost a fifth of the American wage) has the usual water and sanitary problems for its poor, but computing problems arrive by satellite at the end of the American working day and the answers are received by e-mail at breakfast time in the US.

A variety of bodies, ranging from very old charitable groups to Oxfam, which has gathered vast expertise in the last fifty years, have been helping poor communities, whether villagers or urban squatters, to make small improvements to their water supply. In the world of international aid institutions they are known as NGOs (non-governmental organizations) and a certain grudging reverence is paid to their ability to become the catalysts of local aspirations in ways which elude the ruling élites who know very well how to channel funds into their own pockets and those of large-scale engineering firms.

When you talk to the people actually involved in promoting little local projects in water supply and sanitation, the immediate impression you gain is of their humility. The first thing they are conscious of is that if they had known more they would have adopted a different style of approach to local communities. The second is that they are aware of the limitations of their role. Employed on short-term contracts to dispense appropriate technical aid, it is not their 'business' to agitate against the concentration of local power, including power over water resources, which may be in the hands of landowners or money-lenders, nor to make obvious criticisms of the sexual division of labour. But, needless to say, they frequently find that the allies they need to find in order to make small incremental improvements in any community's access to water are the women whose daily tasks, both as household managers and growers, depend upon finding, transporting and disposing of water (see Chapter 7).

Human beings have lived on the planet for a very long time, and have adapted their diet and their patterns of settlement, fixed or nomadic, to the seasonal scarcity or over-abundance of water. Often earlier inhabitants developed elaborate systems of water management. Western Europe, for example, has a plenitude of water engineering projects, viaducts and canals, built with astonishing accuracy by Roman engineers with slave labour, but neglected and allowed to fall into disuse by subsequent populations. Spain is full of reminders of the skill of the Moorish occupiers in water management, some of it neglected, but parts of the legacy remain in active use to this day. Latin America has a similar inheritance, taken over from earlier civilizations by the Spanish conquerors.

The same is true of Africa and Asia. It is part of the conventional wisdom of Australia that in territories where the lost white explorer would inevitably die of hunger and thirst, the Native Australians would know from inherited wisdom where in the desert life-sustaining sources of water could be found.

The rain that falls so unevenly over the globe results in 'run-off' to the seas and oceans through streams and rivers. A far greater quantity is simply evaporated either directly or released into the atmosphere by foliage. Water-harvesting is the name given to human efforts to reduce the quantity lost by evaporation and increase the run-off, using it for agricultural and domestic purposes.

In many parts of the world, in the Middle East and North Africa, India, China, the North-West of Mexico and the South-West of the United States, evidence has been found that earlier inhabitants in dry climates practised 'run-off agriculture' by building *bunds* or embankments to direct water towards their fields of crops, or by constructing terraces on hillsides to conserve rainwater and reduce soil erosion. In India the use of small reservoirs known as tanks for the same purpose is famous. (See the views of Patrick Geddes on the importance of these tanks, cited in Chapter 4.)

Gathering and storing rainwater from rooftops and paved yards is a tradition all over the world. Local regulations compel householders in Bermuda to do so, and this is also true of Gibraltar, where every visitor notices the huge artificial catchment area (35 acres, or 14 hectares) constructed of corrugated iron sheeting on one side of the Rock. But there are many human societies where this material is an impossible luxury and where the ubiquitous old oil drum as a water container is equally unavailable. Part of the task of the development agencies is to spread around the techniques of improvising both guttering and pipes and containers from local materials and local skills, including methods for separating the initial *first flush* or *foul flush* in places with a long dry season, from subsequent potable water.

The most famous example of the rediscovery of ancient methods of run-off farming has been in the Negev desert in southern Israel. Robin Clarke explains that

> *Three experimental farms there are devoted to research and development of water harvesting techniques. Two of the farms – the Avdat and Shivta farms – are based on reconstructions of ancient farms first developed by the Nabateans some 2400 to 1500 years ago. The third farm was founded later, in 1976, to demonstrate the applicability of ancient systems to modern needs. Over the past 30*

years this research has succeeded in growing, in the desert, crops such as olives, almonds, apricots, peaches, pistachio, pomegranates and grapes, wheat, barley, sunflowers and a number of fodder crops.[3]

Clarke, like every other observer of rainwater harvesting, warns against the assumption that the Negev techniques can work everywhere. They have been transported to places like the Khost Plain in Afghanistan, which shares the same pattern of winter rainfall. In other regions, other techniques of water collection have proved more successful. The whole field is examined, with fascinating technical detail, in the standard account from ITDG, by Arnold Pacey and Adrian Cullis, *Rainwater Harvesting*.[4] They participated in and reported on water-harvesting projects in several continents. Most of their instigators operate on the assumption that local community-control of water-sharing installations is the only guarantee of fair sharing and of adequate maintenance. These authors do not contradict this view, but they do point to the difficulties of applying it, especially for the people hired to pass on techniques rather than ideologies.

For example, in the hinterland of Bombay, a body called the Gram Gourav Pratisthan (Village Pride Trust) has sought to bring together households possessing tiny plots of land to co-operate in managing a water resource for irrigation:

> *This makes it possible to grow two crops each year instead of one, and the results have been sufficiently impressive for recent migrants to Bombay with land rights in the area to join the water management groups and subscribe their share of the cost. In one village, some 25 families out of 100 were living in Bombay, and the project leader estimated that 15 would shortly return 'for a better life than in the slums'.*[5]

But these authors are also careful to point out that in several instances in South India, for 'historical and economic reasons, the sense of being one village community does not exist', and they add that:

> *The importance of effective organization is also emphasized by the fate of many traditional rainwater or floodwater harvesting systems under modern conditions. Often the labour to build and maintain*

dams, tanks and run-off collecting channels was organized under a degree of pressure from authoritarian social structures. Sometimes these were relatively paternalistic and fostered a sense of village solidarity and co-operation; sometimes, as in parts of West Africa (as well as Asia), water resources were developed under systems of caste-work and even slavery. With the decline of such institutions, maintenance of old earthworks has often been neglected and the construction of new ones has ceased.[6]

Insistence on local control and maintenance of water installations is an ideological matter as much as a practical one. Some critics lampoon the aid agencies as people who in their own countries are content to take their water supply from public or private agencies and pay the bills without interfering with its management, but who demand that poor people in poor countries, whose lives are spent in physical work to stay alive, should busy themselves in running and maintaining their own water supply. When Duncan Miller of the OECD Development Centre examined the evidence, he found that two key points emerged:

One, where experience is lacking, begin with modest objectives; address relatively small groups; and seek to utilise indigenous traditions, role associations and existing institutional arrangements. This need not mean project planners are doomed to working only with local elites.

Two, self-help and popular participation are not readily transferable; they are effective where there is a spirit of co-operation between the centre and local communities and a tradition of some degree of decentralisation. Whereas partial and paternalistic attempts to mobilise local resources usually do fail in method and objective, the advantages of broader-based participation can be reaped over time and space.[7]

Lack of access to money and advanced technology does not imply an absence of organizational sophistication. This is evident in the accounts of traditional forest and water management in Nepal and Bali cited in Chapter 2, and is shown in an account, also from Nepal, by Tarak Badahur, who comes from the Banspote village on the Sewar river, which uses a series of canals (*kula*) dug into the hillside. Farmers, large and small, operate a 'fair shares' system, which, as in the Spanish example cited earlier, recognizes the difference between access for farmers at the top end and the tail end of the system:

> *in times of scarcity, when there is little water in the* kula *or it does not seem enough to meet the needs of the irrigators, the members of the system gather together and allocate the days and areas to use water on a rotation basis, according to the needs of an area, its water supply and the conventions of the system. During the allocated time, another area cannot use the water, except its seepage . . . until its turn comes. For this purpose the whole area . . . is divided into different parts on the basis of villages and location of fields; and according to the size of the fields of a particular area the rotation schedule is worked out . . . Sometimes, when there is too little water in the kula, then even the fields of one area are irrigated in turns. Such distribution continues until the water volume increases in the kula. If there is a good monsoon the need for rotation does not occur. However, after weeding, or from late September, such an arrangement is made until the time of ripening of the paddy.*[8]

There is no need for imported expertise to tell these people how to organize their water supply, based, as Donald Curtis observes, on universally understood principles. 'When asked, people will say that equal access to water is a matter of principle and that nobody can be denied. However it is also clear that this only applies to those who contribute labour.'[9] But this does not mean that they could not benefit from the sharing of techniques of water management or of reducing salination or silting that can be passed on from other parts of the world.

The range of unofficial aid organizations, raising money in one country and spending it in another, learn from the experience. The Canadian writer George Woodcock was a founder of Canada India Village Aid, which he saw as 'a small and successful example of libertarian organisation'[10] which made contact with people with similar ideas in India.

> *Operating from Udaipur and working among tribal peoples, Seva Mandir strengthened our belief in an approach based on helping the people to pick their own goals and helping them achieve them; there was nudging, shall we say, but not shoving . . . When a drought*

> began in the areas of Rajasthan where Seva Mandir operated, we expanded into the environmental area, forming a partnership between Seva Mandir, which provided the technical services, the villagers who offered their labour, the Indian government which opened its granaries to compensate them, and we who provided the cash for the stone and cement and transporting it. We built ten dams, each of which served a thousand people and their animals as it filled with ground water and the occasional rain.[11]

There are plenty of water-starved parts of the world where the purchase and importation of cement for the building of *bunds* or small dams would be an impossible luxury. The Sahel is the name given to the vast area of Africa south of the Sahara, stretching from Senegal to Somalia, and crossing the frontiers of several nation-states, which is one of the world's poorest economies, whose situation has been steadily worsened by periodic droughts, by famine, warfare and banditry (using modern guns bought by governments from the international arms traders), as well as by outbreaks of contagious diseases among herds of domesticated animals.

In some parts of this vast region international agencies have urged the growing of groundnuts, haricot beans and cotton for the world market, to the neglect of subsistence crops for local use.[12] Others include the Turkana people in an arid region in the northwest corner of Kenya. Their livelihood has evolved as nomadic pastoralism,

> evolved as a lifestyle adapted to the problems of coping with sparse and erratic rainfall . . . They depend on their animals for milk (a major part of the diet), also for meat and blood. They eat wild fruits in season, small game and also some cereals. The latter are bought or bartered or else are grown in sorghum gardens planted mostly by women during the rainy season.[13]

Their plight was seen as an international famine disaster by 1979–80, though it was more likely the result of an epidemic which wiped out large numbers of animals and the fact that, as the European Community investigators found, they were 'particularly vulnerable as a result of heavily armed raiders in Uganda and southern Sudan, many of whom were armed with sophisticated weapons, including automatic rifles'.[14] A series of agencies provided emergency food and relief supplies. They included the European Community, the Netherlands government, Catholic Relief Services, the Red Cross and the Salvation Army. They were later assisted by Oxfam (UK and Ireland), and Adrian Cullis, co-author of the handbook on rainwater-harvesting, was appointed to help introduce a programme of basic water conservation.

He found that what he learned in the Turkana district in the 1980s might have relevance to other places and to other development workers. For example,

> *the assumption behind many famine relief programmes is that once immediate needs are met, a rehabilitation or development programme will be necessary to help people restructure 'outmoded' ways of living to make them more productive. Seen at its worst, this has included the widely-held belief that nomadic pastoralists can only join the modern world if they cease their wandering lifestyle and become 'settled'. The purpose of such measures is usually conceived as making poor people richer, but wholesale destruction of their traditional institutions may well leave them less confident and with fewer coping mechanisms for dealing with adversity.*[15]

The account he has provided, with Arnold Pacey, of the adaptation of water-harvesting techniques to local control in this region bounded by the frontiers of Sudan and Ethiopia to the north, Uganda to the west, and Lake Turkana to the east, is full of detail of practical experimentation as well as social accommodation and the determination to leave a workable system in local hands.

Patrick Mulvany, agricultural adviser to the ITDG, urges us to take note of the fact that their book closes one chapter and opens another in the history of intermediate or alternative technology:

> *The previous emphasis on technology transfer is giving way to a participatory technology development process in which the producers set the development agenda. Before, agency ideas were often imposed. Now, it is the producers' voice that is dominant. Control of work, initiated by the agencies, has passed from expatriate project workers to the pastoralists represented collectively in a management committee. In this changeover no point was more critical than the decisions taken at the meeting about the management of the project after the expatriates left . . . Unless the agencies wish to perpetuate the*

iniquities of the old order, all of us in the development agencies must seek ways of working to the new agenda that is set by the producers: we must take sides.[16]

The most obvious of providers of water, all over the world, are women. For everywhere in the world they are the managers and usually the carriers of the water used for every purpose in the household, and very often in horticulture and animal management too. But they are frequently excluded from decision-making about water supply. This is all too obvious in poor countries, but is equally true in rich societies where water is available from household taps.

When non-payment results in the cut-off of a household supply, as in Britain in 1996, who in the household has to go with a bucket to a neighbour to beg for water for this basic need? You can readily assume that it is a mother, seeking a minimum for her children's needs; just as in the poorest parts of the world, water is women's work.

Notes

1. E. F. Schumacher, *Small is Beautiful* (London: Blond & Briggs, 1973), p. 165.

2. George McRobie, *Small is Possible* (London: Jonathan Cape, 1981), p. 121.

3. Robin Clarke, *Water: The International Crisis* (London: Earthscan Publications, 1991), p. 147.

4. Arnold Pacey and Adrian Cullis, *Rainwater Harvesting: The Collection of Rainfall and Runoff in Rural Areas* (London: Intermediate Technology Publications, 1986, 1989, 1991).

5. *Ibid.*, p. 38.

6. *Ibid.*, p. 41.

7. Duncan Miller, *Self-Help and Popular Participation in Rural Water Systems* (Paris: OECD Development Centre, 1979), p. 36.

8. Tarak Badahur KC, *Farmer Managed Irrigations Systems in Nepal, A Case Study* (Birmingham: Dept of Development Administration, University of Birmingham, 1986), cited in Donald Curtis, *Beyond Government: Organisations for Common Benefit* (London: Macmillan, 1991), p. 74.

9. Curtis, *op. cit.*, p. 75.

10. George Woodcock, *Walking Through the Valley* (Toronto: ECW Press, 1994), p. 156.

11. *Ibid.*, p. 155.

12. Nigel Cross, *The Sahel: The People's Right to Development* (London: Minority Rights Group, 1990).

13. Adrian Cullis and Arnold Pacey, *A Development Dialogue: Rainwater Harvesting in Turkana* (London: Intermediate Technology Publications, 1992), p. 1.

14. *Ibid.*, p. 8.

15. *Ibid.*, p. 120.

16. Patrick Mulvany, Preface to Cullis and Pacey, *ibid.*, p. vii.

CHAPTER 7

The Women at the Well

> With these facilities we escape the drudgery which households have to endure in poor communities across the world where the water they need has to be fetched and carried home each day, often from sources several miles away. This drudgery falls largely on women, as part of the housekeeping task. Once the water is put in pipes the work of managing the distribution network is called engineering and becomes much more of a male operation . . .
>
> **David Kinnersley, *Troubled Water*[1]**

In a small Rhineland cathedral town called Xanten, I paused to admire the elaborately decorated cast-iron water pump with its wrought-iron handle. A resident explained to me, 'You will find a pump like this in every street, and you really ought to be here for the Peter and Paul day festival when the whole community gathers around its pumps to celebrate the gift of water.'

Sure enough, in another street I found that the exclusion of traffic had enabled the town to commission a series of life-size bronze figures of people waiting to draw water from the pump. Apart from the children around their skirts, they were, of course, all women. It is a theme with resonances all over the world. In the 1950s, when the wave of self-criticism among architects and town-planners began to rise, a favourite anecdote spread around by the architect Constantinos Doxiadis concerned the women of a North African town who met at the well to collect water or at the river to wash the family's clothes. When water became available from a tap in the home or from a standpipe in the courtyard, these daily communal contacts were lost.

Now I doubt if, after childhood, either Doxiadis or Le Corbusier, who also related this history, had ever been obliged to draw water from the well or to play a part in the routines of endless laundering. Nor have I ever understood the precise message that the story was intended to convey. Were architects and engineers to refrain from replacing the drudgery of daily life for fear of disturbing traditional patterns of female seclusion? Was it wrong to liberate women from servitude to back-breaking labour?

All the same, it is true that, as we saw in a British context in Chapter 1, the daily work of collecting water has always been one of the agencies for cementing social bonds on the basis of shared common needs. It is also true that in the sexual division of labour throughout the world, women, as household managers, are expected to be the procurers of water and to manage the quantity required for the needs of the family for feeding, personal and domestic cleanliness and sanitation, child-rearing and family laundry. In many cultures they are also the horticulturalists, responsible for keeping food crops alive. To them too falls the task of toilet training for children, protecting them from water-borne disease, maintaining family health. All these tasks depend on water, and on continual assessment of its suitability and safety.

Yet, at the same time, most societies have a tradition that the technology of water supply is a matter too complex for women. As David Kinnersley explained, it 'is called engineering and becomes much more of a male operation'.[2] A story from West Africa illustrates the way in which the different social roles of the sexes distort decision-making about water:

> At a village meeting in Bangu, East Mamprusi District, discussions on the siting of a well were nearing completion. The meeting had decided on a site in a valley. Then, almost as an afterthought, the extension worker asked the women for their views. Gradually they spoke up and pointed out that the men had chosen the valley site because they knew it was going to be easier for them to dig a well at the valley site than at an alternative, but nearer, site. If the well were dug in the valley, women would have to carry water uphill every day, the journey from the other site would be shorter and less steep.[3]

In this instance the well was dug at the nearer site, with more short-term work for the men, but a lesser daily burden for the women. A precisely similar story was told me by Geoff Sands, a builder employed by a relief agency in El Salvador, a country ruined by a decade of civil war. The local community leadership agreed with the proposal of a northern agency to fund

> the site works for the installation of a hydraulic ram pump — work for the all-male building team but otherwise pointless and now dismantled — at the same time as the women in the community were having to walk long distances to collect water from a defective standpipe system.[4]

There is a body in the Netherlands called the International Reference Centre for Community Water Supply and Sanitation, and one of its workers, Christine van Wijk-Sijbesma, undertook the formidable task of collating nearly 800 reports from all over the world. Her analysis is an outstanding testimony, not merely to the handicaps that women face, but to their endurance and ingenuity in doing their utmost, not for their individual needs, but for those of the family.

On the issue of man as engineer and woman as water-carrier and sanitary orderly, she cites the importance of woman professionals as agents of women's priorities in the management of water:

> Women in a Karachi squatter area responded to the efforts of a concerned woman architect-planner to get them to undertake the improvement of their own community. Despite male community worker statements that only men did that sort of thing, the inspired women harnessed the efforts of their out-of-school adolescent boys to carry rock fill from a nearby hill, and dredge the stagnant canals in the neighbourhood. The women were right there directing the work all the way.[5]

But there are far less spectacular ways in which women are the carriers of water wisdom. They have the melancholy knowledge of which children died and which survived. Women, all around the world, decide how to use any particular source of water at any time of the year, and through experience they 'make reasoned decisions based on their own criteria of access, time, effort, water quantity, quality and reliability'.[6] The World Health Organization continually reminds us that 80 per cent of all disease in the 'developing countries' relates directly to unsafe drinking water and lack of adequate hygiene. Women do not choose to provide the family with contaminated water, to tolerate inadequate and dangerous latrines for the men, and still less satisfactory provision for themselves, nor to allow the children to defecate all around the home, and to have no means of cleaning themselves.

This is the climate of deprivation throughout the world of the poor, including that of families in the rich world, where the application of market forces to water supply has imposed on people who know perfectly well about basic hygienic precautions a climate of dirt, improvisation and furtive faeces disposal.

The worldwide survey continually stresses this fact:

There is a great deal of evidence that in all cultures, women, through their daily experience and observation, have acquired basic and practical knowledge of environmental hygiene on which participatory programmes can build. Reference has already been made to their traditional practices of source selection, in which they make reasoned choices and often distinguish water quality according to use and to the characteristics of the source. An exception in this respect is tap water, which frequently is considered to be safe when it looks clean, even if it comes directly from a river without treatment. Projects should not keep users in ignorance of such issues, but need to discuss them as part of local decision-making on the choice of technology . . . [7]

I have often heard criticism of the work of aid agencies, whether unofficial bodies (NGOs) or governmentally financed bodies, that they move into an area with a package of promises of expertise, but arrogantly ignore the traditional channels of communication by way of dominant chiefs or headmen, local landowners and money-lenders, in a way that the one-time colonial administrators would never have done. The evidence contradicts this assumption. The successors to old empires have been not only a series of territorial wars, and the economic imperialism of cash-crop production for the export market, but also a range of vast engineering projects which are notoriously damaging to local populations.

The unofficial aid workers, even when funded by national or international government agencies, have learned through experience that to change the situation anywhere, as the previous chapter shows, they need to win the support of the humblest of local people, and that these are often the women.

Unlike the employees of international conglomerates involved in enormous water-engineering projects, they, while hired to propagate simple techniques of local water-management, have to be concerned with winning agreement and support. They have also to ensure that there are simple and acceptable routines for day-to-day management, without dependence on spare parts for which there exist no budget or ordering procedures accessible to villagers. With no brief to intervene in the existing structure of power, influence, and access to land, they have to do what they can within the cultures and societies that have sought their aid.

They rapidly learn that their allies are the women, who daily carry the burden of household management and, in many societies, of horticulture too.

In the rich world it is impossible to convey the liberation implied, not by the dream of the tap in the home, but by the pump or the standpipe in the yard or street. The worldwide survey makes this abundantly clear:

> *The health of women and their families will also benefit from reduction in time and energy spent on water collection . . . Women need to expend less energy from their often low food intake on the heavy task of transporting water, so that they have more energy for themselves, for foetal development and breast-feeding, and more time and vigour for tasks essential to family health, including household hygiene, child care, production and processing of food crops, cooking and income generation. Water collection is not only energy consuming, but may also have detrimental physical consequences. Carrying heavy water pots, for instance, is mentioned as a primary cause of pelvic distortion, which may in turn lead to death in child-birth. In villages in dry areas in Thailand, water collection is one of the seasonal stresses reflected in data on miscarriages. The risk is high of falling on slippery paths and steep slopes, while carrying food, water, and a baby. Half of the cases of broken backs treated in a rehabilitation clinic in Bangladesh resulted from falls while carrying heavy loads.*[8]

But, as we have seen, decision-makers often have other priorities. An account of a Dutch-sponsored water project in

Guinea-Bissau stressed that the main water-users, the women, had to walk long distances to reach unsatisfactory wells, 'but they could be approached only with the consent of the men' for whom drinking water was not the greatest priority, as they were more concerned with irrigation water; so, when a new well site was chosen, 'in many cases the access road was not built because this was men's work, whereas collecting water was women's'.[9] Similarly close familiarity with disease, death and disablement in the family enables women, rather than men, to distinguish between the origins of different kinds of illness in relation to water, which can be classified in four categories:

Water-borne: water carries the infection, such as typhoid and cholera.

Water-washed: lack of washing affects skin or eyes, as in scabies or trachoma.

Water-based: via parasitic worms depending on aquatic life-cycles, as in schistosomiasis and guinea worm.

Water-related insect vectors: such as malaria and yellow fever.[10]

They all call for different precautions and remedies, most of which are beyond the reach of most of the world's women. As David Kinnersley notes, 'Lack of water for washing especially promotes diarrhoea, a great killer of children under age five, as so much of it arises from handling food with unwashed hands.'[11] Similarly, the elementary precaution of boiling water, so easy in the Western world, is surrounded by difficulties in the poor world, because of the scarcity of fuel and the inefficiency of stoves. Patricia Adams and Lawrence Solomon, reporting on the rich world's assumption that the poor world wastes wood fuel, comment that

The experts' error lay with their assumption that they understood what the Third World needed – an assumption based on nothing more than a conviction that Western know-how, which could put a man on the moon, would be capable of solving something as simple as the Third World's cooking fuel shortage. In the experts' enthusiasm to produce the product they assumed the Third World had been waiting for, they neglected to conduct the basic market research that would be commonplace in the West – for example, asking the women who would be using the stoves whether they suited their needs.[12]

In the poor world people do not buy a cast-iron, energy-efficient stove, but a cheap one, made locally out of scrap metal, usually old oil drums, and suffer its deficiencies. When I first met the architect Madhu Sarin, she was championing the cause of the unofficial economy in Chandigarh, the new capital of the Punjab, but I learned that one day she chanced to visit a Harijan hamlet, Nada, 15 km from the city. Meeting a woman who was literally in tears from the smoke of her stove, which blackened the walls, ceiling and furnishings, she set about developing an improvement to the stove (*chulha*), introducing a damper system which was cheap and simple and eliminated smoke. The result was not only that 'villagers now collect firewood only once a week rather than twice a week', but that women 'have taken to the new *chulha* not because it is efficient but because it is smokeless'. And it is reported that the Sarin stove seems to be spreading to other parts of India, where 'the "experts" going into the communities are not men but women volunteers . . . women find it easier to communicate with the women.' And Madhu Sarin herself comments that her chance success was the result of 'understanding the villagers, their problems and perspectives . . . unless one identifies with the people by helping to solve their problems, the villagers will not listen to any advice, leave alone working on it.'[13]

This small solution to a specific domestic difficulty brings one particular water-related issue within the control of the household. Others depend on community endeavour, in which the public role of women is vital. David Kinnersley, who convened the 1981 Thirsty Third World conference and initiated the voluntary organization WaterAid, believes that, precisely because women across the world are not only the fetchers and carriers of water, but its main users in the household and horticultural economies, theirs is the missing voice:

> *The need is to give women far more scope at local level to lead and influence community initiatives, in the shaping of services that families want, as well as in business. To achieve this – throughout South America, Africa and Asia – longer schooling for women is a necessary and urgent step. Nor must the pressures of increased population in the mega-cities swamp and frustrate this aim. Given more chances to develop their potential, women can moderate those*

pressures in direct ways. The moves to involve more citizens in consultation, more involvement of non-governmental organisations, and the sheer wastefulness of excluding women must surely all bring much more change on this front in future. The water sector could provide one of the best springboards for such change.[14]

But in liberated Britain, with legislation to outlaw sexual discrimination, and where housing policy has concentrated the jobless poor in municipal estates where until 1989 water charges were usually included in the rent, it is usually women who have to cope with the consequences of non-payment of water bills and with the threat of disconnection and the prospect of raising children in a household deprived of water.

Notes

1. David Kinnersley, *Troubled Water* (London: Hilary Shipman, 1988), p. 172.

2. Ibid.

3. Jan Davis and Gerry Garvey, *Developing and Managing Community Water Supplies* (Oxford: Oxfam, 1993), p. 67.

4. Geoff Sands, 'Reconstruction in El Salvador: The social relations of community self-build', forthcoming in the Oxfam series *Development in Practice*.

5. Christine van Wijk-Sijbesma, *Participation of Women in Water Supply and Sanitation* (The Hague: International Reference Centre for Community Water Supply and Sanitation, Technical Paper 22, 1985), p. 116.

6. Ibid., p. 89.

7. Ibid., p. 36.

8. Ibid., p. 15.

9. Sumi Krishna Chauhan, *Who Puts the Water in the Taps?: Community Participation in Third World Drinking Water, Sanitation and Health* (London: Earthscan Publications, 1983), p. 60.

10. David Kinnersley, *Coming Clean: The Politics of Water and the Environment* (Harmondsworth: Penguin Books, 1994), p. 19, citing G. White, D. Bradley and A. White, *Drawers of Water* (University of Chicago Press, 1972).

11. Ibid., p. 189.

12. Patricia Adams and Lawrence Solomon, *In the Name of Progress: The Underside of Foreign Aid* (London: Earthscan Publications, 1985), p. 82.

13. Ibid., p. 130.

14. Kinnersley, ibid., p. 189.

CHAPTER 8

Marketizing Water

> *All proclamations that water is 'an economic good', that it must not be gratuitous and that economic pricing must regulate access to it, stem from a fundamental confusion between the limitedness of water and its scarcity . . . If access to water were to be governed by the law of scarcity, its price would skyrocket in such a way that the poor would not get a drop of it. Even if it seems wise to impose high water tariffs on industries, agribusiness and private over-consumers, the poor's access to gratuitous water must be guaranteed. This claim is nothing else than the modern expression of the unwritten law of traditional societies: protect the subsistence of the weakest . . . Before any debate on the style of its pricing, water must be recognised as what it has always been: a commons.*
>
> Jean Robert, *Water is a Commons*[1]

We saw in Chapter 1 that the Victorians, despite their faith in the economics of the market, recognized a 'binding moral duty' to ensure that every household had access to a clean water supply, regardless of their ability to pay, but that in 1994, 12,500 English households had their water supply disconnected through non-payment of water bills.

In 1887 the anarchist propagandist Peter Kropotkin saw 'water supplied to private dwellings, with a growing tendency towards disregarding the exact amount of it used by the individual', as one of the signs – along with free roads, free libraries, free public schools, parks and paved and lighted streets for everyone's use – of a trend towards a society where 'everybody, contributing for the common well-being to the full extent of his capacities, shall enjoy also from the common stock of society to the fullest possible extent of his needs'.[2]

A century later acceptable opinion had made huge shifts, for two reasons. The first was the rise of the revived religion of the market and the privatization of public goods at any cost. The second was the growth of ecological consciousness and the realization that all resources are finite. For example, Sandra Postel, a renowned authority on water scarcity, remarks that,

> *Amazingly, water charges for most British households are linked to the value of their home, and have nothing to do with actual consumption . . . Trials in the United Kingdom have shown that metering can cut household use there by 10–15 per cent.*[3]

And in 1995, when the British government issued its report *Water Conservation: Government Action*, the emphasis was on charging according to consumption. An opposition spokesman complained that 'Twenty-nine paragraphs of this 71 paragraph document are devoted to water metering. The Conservatives want to force everybody to have a water meter.' And he added that figures from within the water industry showed that 826 million gallons of water were lost every day from leaking water company pipes which would not be affected by metering.

> *The government themselves admit that it would cost up to £200 per household to install water meters. That would cost the customers between £4 billion and £5 billion. Meters would also cost up to £500 million a year to run.*[4]

Water conservation is a vital issue in rich and poor countries alike, but is in fact trivialized by the application of the price mechanism to rich and poor alike. Thus Jean Robert comments that

> *Governments intending to rule water consumption by price mechanisms should remember that only as long as water is a commons, freely accessible to the poor, can the over-consumption of the rich be curbed by high tariffs without causing the poor's ruin. In Lima, for instance, where the government has tried to regulate the use of water by that means, prices are too high for the poor, who buy*

water by the drum, and too low for the rich, who bribe the drivers of tank trucks giving service to the slums and use the poor's water to wash their cars.[5]

He struggles with the issue of the conflict between equity and efficiency in urban water supply. Others, like Jim Amos, Senior Fellow in Birmingham's Institute of Local Government Studies, reject what they see as 'the fallacious reasoning that water is a natural resource and therefore should be freely available', but also seek to avoid the situation where the poor have to pay far more to buy their water from vendors than do those who are connected to the mains supply, for:

There is also the rather more humanitarian argument that water is such an essential of human life that the poor should not be deprived of a supply of water because of their poverty. Consequently many water supply systems were developed by local authorities or other bodies with access to public funds, thus making it possible for water charges to be subsidised. Since the managing institutions have frequently not been obliged to recover costs directly from consumers, there has been a tendency for the subsidy to increase at least in gross terms. Now, faced with the need to make major capital investments in order to meet the increased need/demand, water authorities have difficulty in securing development finance.[6]

Amos explains that to overcome the problem of development finance, many water supply bodies have been established in recent years 'as separate entities required to apply charging systems which will, in the long term, convert them into viable commercial operations'. He adds: 'However, it is not uncommon for strong political and consumer pressures to inhibit this transition.'[7]

In Britain the privatization of the water supply industry was uninhibited by effective opposition at the time, and its consequences were not publicly noticed until several years later. The administrative detail on the British water industry that I provided in Chapter 1 is important for two reasons. The first is that the reforms of 1974 brought the supply and disposal of water, for the first time, under the direct control of central government; and the Treasury (the finance ministry), under both Labour and Conservative control, steadily reduced the spending of the water authorities between 1974 and 1986. By 1982 the government was permitting the water industry to spend only half the sum it put into capital investment in 1974. The second reason is that the government's tidying-up of the water industry enabled it to join the other publicly owned utilities offered for sale in the 1980s to a public which already owned them collectively.

The British Isles are water-rich, with adequate rainfall, but occasionally experiences a water shortage. It is fascinating to compare public attitudes to drought conditions in 1995 with those that emerged in the sixteen dry months between May 1975 and August 1976. British households in 1976 were unaware of future shifts in perception that changed water from a common good to a commercial product. Fred Pearce, the water correspondent of *New Scientist*, reported how until then water-planners had seen 'any form of supply restriction – even a hosepipe ban – as an admission of their failure. They realised that it was nonsense to spend millions to cut the frequency of hosepipe bans from one year in five to one year in ten.'[8] Yet he also learned about the improvizations of the then regional water authorities in pooling access to water sources, as well as their long-term plans like the London ring main and the device of recharging aquifers from winter river water.

The public response was even more interesting. The National Water Council found that

> *the potential for voluntary savings by the public and industry during a water crisis was vast. The 'save it' publicity campaign during the drought cut water demand by 30 per cent in some areas . . . and further measures like emergency leak detection and pressure reductions in the mains made savings of another 10 per cent.*[9]

Ninety per cent of the population answered pleas to cut down on bathwater, and over 80 per cent said that they had taken more care to put the plug in basins and sinks, though only 9 per cent said that they had put a brick in the WC cistern.[10] In 1976 there was intense co-operation between water authorities, so that Fred Pearce notes that 'in the worst drought for 250 years, engineers managed to keep water flowing'.[11] There was active co-operation from the public in reducing demand. There were no recriminations: simply a willingness to learn from the experience.

By the time of the drought of 1995, the climate had changed. The public placed the blame on the water companies and the companies blamed the public. The Secretary of State for the Environment, John Gummer, advised people to follow the precepts of 1976 and recycle washing-up water on the garden and put a brick in the WC cistern. My local newspaper, which is hardly a radical journal, pointed in a leading article to the crucial difference between then and now:

> *But then water was public property, and the public had an interest in conserving it. We have since been re-educated by Mr Gummer and his Cabinet colleagues, to think of water not as a natural resource, but as a capitalist product.*
>
> *Newly-privatised water companies have sought to justify exorbitant profits by telling us what a vastly improved service they are providing. They have rewarded themselves with enormous 'performance-related' pay rises. Are we not entitled to expect that, as long as we pay our bills, we should be able to use just as much of the stuff as we like? And should it matter to the profit-centred water companies whether we choose to use it for watering our gardens or flushing the loo? Isn't the buyer entitled to use it as he likes – just like any other commercial product?*
>
> *Of course, such attitudes do not fit well with conservation, but if conservation had been properly considered at the time, perhaps privatisation would not have seemed such a good idea.*[12]

The attitudes, both of water-suppliers and water-users, in the droughts of 1976 and 1995, reflect two opposed concepts of what it means to live in a civilized society. When the government sold, at a loss, the water supply installations and infrastructure of England and Wales, there was no discussion at all of the problems of poor people unable to pay for their supplies. Mounting criticism of policies of disconnection have led some of them to devote a minute proportion of their pre-tax profits to charitable trusts to help some of their poorest, most indebted customers.

> *Three water companies, attempting a metamorphosis from Scrooge to Santa, say they have set up the charities, or are in the process of doing so, because some customers are too poor ever to get out of debt . . . The state benefit system will not help these debtors and they have no*

> *prospect of earning enough to break free. There are advantages for the companies beyond mere public relations. The debt pay-offs help them reduce the number of disconnections . . . There is also a tax advantage, although the sums involved are trivial compared with the profits the firms make and the corporation tax they pay.*[13]

As we have seen, in the latter half of the nineteenth century, faced with the fact that private water companies had no interest in providing a water supply to poor households, local authorities were obliged to buy the water companies to make provision universal. In the late twentieth century, central government assumed control, and by the end of the century, with its enthusiasm for market forces, 'freed' the water industry, which belatedly recognized the poor as objects of charity. We have witnessed a frightening decline in public acceptance of the concept of water as a common good.

But despite their tarnished reputation at home, the British water companies have, like their French counterparts, roved the cities of the world for contracts to undertake water supply and sewage disposal. Thames Water, which told the *Guardian*'s reporter that 'We are being too soft, and that is why our disconnection levels will rise', is part of a consortium reorganizing the water supply of East Berlin, where domestic water under the pre-1979 regime was provided without charge.[14] Suffolk Water and Essex Water are both owned by the French company Lyonnaise des Eaux, which, with the larger Compagnie Générale des Eaux, are the world's leading water distributors. Thames announced in 1994 that it was teaming up with Générale des Eaux to bid for the contract to run, maintain and expand the water and sewage system of the Peruvian capital Lima,[15] while Anglian Water in 1993 was part of a consortium led by Lyonnaise des Eaux to modernize and run the water and sewage operations of the Argentine capital Buenos Aires.[16]

It would be comforting to imagine that European technical expertise and financial probity were being exported to benefit the poor of Latin America, but when the chief executive of Thames Water was asked about its acquisition of an American water engineering subsidiary which was losing money, he explained to the press that 'The aim remains to expand non-utility businesses to provide an earnings stream free of regulatory control.'[17]

Marketizing Water

In Buenos Aires the government will continue to own the water undertaking, *Obras Sanitarias de la Nacion* (which fails to provide water to over 3 million people in the area), but the consortium will bring its own expertise, and will 'manage the capital investment programme with funds generated from revenue and normal commercial borrowing'.[18] The geographer Erik Swyngedouw, in a study of the contradictions of water supply in Guayaquil,[19] the largest city in Ecuador, throws light on the dilemmas of all the expanding cities of Latin America. Urban politics have a long tradition which he describes as 'clientelist-populist', which result in affluent areas of the city having a heavily subsidized water supply built on large-scale nineteenth-century engineering. 'From the very beginning, the high cost of urban engineering works required massive (usually external) financing, while the political-economic forces in the city demanded low water prices. This results in chronic losses for the water utilities.'[20] With some notable exceptions, the majority of Latin American cities

> *face the fact that a sizeable part of the urban population – and invariably the poorer end of the social spectrum – does not have access to piped potable water. This situation, in turn, makes them easy victims of water speculators, i.e. the private water sellers who distribute water in non-serviced areas by means of tankers. In Guayaquil, for example, approximately 400 tankers service a population of 600,000 (about 35 per cent of the total urban population). These water merchants buy water at a highly subsidised price, while . . . the price they charge is up to 400 times larger than that paid by low-volume consumers who receive water from the public water utility.*[21]

This city, built at the confluence of two rivers to form the Guayas estuary, has appalling hygiene and sanitation problems, but no scarcity of water, for Dr Swyngedouw finds that

> *the average production and supply capacity of the existing facilities would allow each inhabitant an average daily consumption of 220 litres. Current consumption varies between an average of 307 litres per inhabitant in the well-to-do parts of the city to less than 25 litres for those supplied by the private water sellers. Compared with an internationally accepted standard of 150 litres per person per day,*

> *Guayaquil is in a position to provide every citizen with a sufficient supply of potable water . . . The water problem is, therefore, not the result of a routinely invoked 'natural' scarcity or of environmental or technical constraints.*[22]

Consequently he rejects the 'productionist logic' of the Anglo-French water experts, which simply adds to the massive debt burden of Latin American economies, which combined with imposed austerity and market-orientated policies invariably imposed by international lending agencies serves to 'divert capital away from social and collective investments'.[23] He explains the background to contemporary water imperialism:

> *The requirements of international lending agencies make project execution subject to international tender. Bilateral loans are usually subject to the use of engineering services and procurement of equipment from the lending country. These mechanisms privilege new infrastructure investment over institutional or organisational improvements and maintenance projects.*[24]

Yet it is precisely institutional and organizational change that the fair sharing of a precious resource demands. We are back with the problem faced by Jean Robert in asserting that water is 'a commons' and not a commodity. The poorest of inhabitants in the unofficial settlements that surround the cities of the South have, in addition to the endless problems of survival, the task of asserting their right to their share of this common good. They cannot rely on the goodwill of politicians unless they become strong enough and organized enough to exercise an influence which political leaders are obliged to placate.[25]

One example will illustrate this truth. It comes from Mar del Plata, a coastal resort city in Argentina, south of Buenos Aires. This fact carries its own irony, since it was the 1977 United Nations Conference at Mar del Plata which resolved that 'All people have the right of access to drinking water in quantity and of quality equal to their basic needs.' This resolution, in turn, led the United Nations to designate 1981–90 as the International Drinking Water Supply and Sanitation Decade, with the aim – which has not, remotely, been realized – of ensuring that everyone on earth had a minimum of safe drinking water.[26]

Mar del Plata has a series of southern *barrios*, or self-built working-class settlements, with an estimated population of about 120,000. Of their many problems, the shortage of potable water was the most sharply felt. Local self-help could not solve this lack for two reasons. The first was a thick layer of rock which frustrated the digging of wells. A community activist, Héctor Woollands, explains that

> To drill as far as water would only be possible with proper equipment which is extremely costly, to such an extent that in some cases putting in a well would be more expensive than building the actual house. In the other part of the area . . . where the thick rock layer disappears and water is easily obtainable, it is highly contaminated, and has been for the last 12 or 15 years, and to drink it or prepare food with it constitutes a grave risk.
>
> This situation has obliged the grass-roots organisations to embark on a very serious search for some solution. Initially they set up taps at 'Sanitary Works' sites located throughout the barrios, but these were never even remotely sufficient. This gave way to private water-sellers. Men with pick-up trucks drove through the barrios selling water, for which they charged more than the price of milk, thus making its purchase impossible for the poorest families. The lack of water for such a large population resulted in drought conditions in the summers and plagued the residents all year round. The protests and demands on the authorities increased as well, and grass-roots organisations began to co-ordinate action in defence of the neighbourhoods. The need for water and the indignation of the barrio dwellers reached such a point that in February 1988 they organised a gathering in front of the Cathedral, where they went to wash their clothes in the 'Peatonal San Martin' fountain. Naturally this irreverent, nonchalant attitude provoked a furious reaction from the municipal authorities, who condemned this conduct. But the real result was that ever since then everyone has known that in the 'happy city' there are citizens who must buy water or go thirsty.[27]

Continuous and sustained pressure, and what Woollands calls *fomentismo* (grassroots-ism), led to the Sanitary Works Department overcoming its dilatoriness in recognizing the problem of water shortage and initiating test drills from which a few of the barrios could be served and at the same time studying the possibility of constructing a Southern Aqueduct and of seeking loans to finance it. The various neighbourhood associations formed a Pro-Aqueduct Commission (CVPAS), with as its fundamental task propaganda and agitation to bring the work into being, but also aiming to organize the distribution of water to residents until then.

In November 1988 it signed an agreement with the authorities for neighbourhood associations to distribute free water in the barrios, with the Sanitary Department providing two tankers and a sum of money to cover the running costs. Wide publicity in Argentina and Uruguay was given to the fact that the Commission distributed 40,000 litres of water every day without charge to the variegated cluster of settlements. Then in 1990, the neighbourhood associations undertook to take charge of the digging and pipe-laying for the new permanent water supply, employing the contractors itself and taking control of finances shared on an equitable basis under voluntary agreements with each householder, payable as a lump sum or in instalments. The costs were found to be half those of similar works carried out elsewhere by commercial enterprises.

Héctor Woollands draws several conclusions from this experience. The first is that joint communities impart lessons and concepts of solidarity and altruism that can only be learned from popular self-organization because 'it provides daily demonstrations of mutual aid, a concept which is not taught elsewhere'. The second is that if responsibility for water supply had been left to the authorities, 'the inhabitants of the marginal barrios of the city would still be thirsty and still paying inflated prices for water of doubtful quality from the *aguateros*'. The third is that the experience is an aspect of the permanent struggle of low-income people against 'the oppressive economics imposed by private companies backed up by the law'.[28]

He characterizes the burden of paying for water with a phrase from the Argentine folk epic *Martín Fierro*, which calls the economic system 'a spider's web that the big bugs can break through and which only ensnares the tiny ones'. As we have seen, this precisely characterizes the Latin American experience of water supply. A study of Lima demonstrates that 'contrary to popular belief, the level of unpaid water use is considerably higher in the central upper-class parts of the city than in the lower-class periphery'.[29] In relation to income, the poor pay more, and the recent history of water changes in Britain illustrates that this is true there too. What has become of the concept that water is a common good?

The Organization for Economic Co-operation and Development (OECD) is an intergovernmental body of 24 industrialized countries. In the 1980s it formed a steering group to advise on economic aspects of water conservation, since 'for both economic and environmental reasons it has become increasingly difficult to meet water demands in their various forms'.[30] It found a great variety of pricing structures for water provision, and classified them into five types of payment for piped services in water supply, sewerage and sewage treatment.[31] They were:

1. *Flat rate tariffs*, where water charges are not directly related to quantities of water used. (This is the method commended by Kropotkin and condemned by Sandra Postel, above.) At the beginning of 1996, 93 per cent of the British population still had an unmeasured supply, but the government, and the Director of Ofwat, the regulatory body, was of the opinion that a measured supply is the best long-term solution to water shortages, cutting consumption by 11 per cent. A report from the Centre for the Study of Regulated Industries showed that the average water charges for people with meters had fallen by nearly 2 per cent in England, while unmetered consumers had witnessed a rise of 39 per cent in water charges. The opposition spokesman, Frank Dobson, claimed that 'the government had a secret agenda to force everybody to install meters'.[32]

2. *Average cost pricing*, where all water service costs, other than access costs, are grouped together and divided among the total number of units expected to be sold, to generate a unit cost. The OECD investigators did not investigate the degree to which private profit-making water-suppliers might enhance the average costs for the benefit of shareholders or directors, nor to their involvement in speculative investment outside the

water industry. This has been the subsequent experience of the sale of the water industry to private monopolies in England and Wales.

3. *Declining block tariffs*, where succeeding blocks of units of water consumed are sold at lower and lower prices. (A fixed or minimum billing charge is usually included.)

4. *Increasing or progressive block tariffs*, where succeeding blocks or units of water are sold at higher and higher prices. The OECD report found that 'these tariffs were becoming increasingly common, and reflected, at least initially, income redistribution objectives (provision of an initial, basic supply at a low rate, with additional consumption becoming progressively more expensive).'[33] This is the approach closest to the view of Jean Robert, quoted at the head of this chapter. The OECD reported, too, that

> *the system was used to promote efficiency and conservation objectives, although it is noted that the evidence of the effectiveness of such tariffs in the industrial sector is not clear, given the concurrent economic recession and restructuring away from large water-using industries.*[34]

This is a bland way of reporting that industrial use of water has declined in the industrialized nations through the collapse of heavy industry.

5. *Two-part tariffs*, including a fixed element, often varying according to some characteristic of the user, and average cost pricing in the form of a single volumetric charge. These too, might be seen as a recognition of a universal right to water: a basic notional fee for the right of access and a subsequent fee for the quantity used, above a minimum.

The OECD report also noted a wide variation in the concept of 'natural' water services, and vast variations in the attitude to irrigation water. The demand for water for irrigation was 'highly responsive to price changes'.[35] Its conclusions were that in order to ensure the most efficient use of a scarce natural resource, the principle of 'marginal cost pricing' should be adopted. The idea that water is a common good became marginal. The investigators found the concept of quality or fairness 'a difficult objective to define' since it ranged from the notion of social pricing, 'whereby no consumer should be prevented by income considerations from enjoying the benefits of water, to narrower concepts such as that provided by the requirement that each consumer should pay the same per unit of water service received regardless of its cost'.[36] Inevitably, across 24 nation states, it found that

> *In some countries, water pricing is structured to foster development in agricultural and industrial sectors, to lighten charges on isolated communities or to help low-income households; in others the view is taken that it is preferable to adopt an economically efficient water pricing system in combination with a social security system for those consumers disadvantaged by the policy. It was thought that any financial subsidies to the water services as a whole, or to one group or water service users from another, should be made explicit and be justified by arguments for special treatment.*[37]

The hasty sale of the water supply industry in England and Wales ignored the need for any special treatment. Not only were regional differences ignored, but water charges, once included in the rent for tenants of public housing, were deliberately excluded from social security payments. The needs of the poor are left to private companies which have chosen to exclude them from a supply of water, but who have belatedly instituted charitable undertakings to rescue those who fit *their* definition of the deserving poor.

About half of the private water companies in England and Wales have selectively introduced, or are testing, prepayment meters obliging users to pay in advance with 'smart cards' which can be recharged on payment at post offices. Customers who fail to pay have their water supply cut off automatically, and the cards ignore the quantity of water actually used, and after disconnection the weekly charges must be made up before the supply will come on. These are not regarded as disconnections, and so are not included in official statistics. It was reported in 1996 that

> *Two thousand have been installed in Birmingham since 1992, and Birmingham council plans to lead an application for a judicial review*

of the water regulator *Ofwat's* advice that the devices are legal. The Association of Metropolitan Authorities has obtained a legal opinion that such meters undermine the statutory regulations governing disconnections. According to Birmingham council, there have been more than 2,000 disconnections in the city since 1992, of which 147 went to court.[38]

We have moved a long way from the concept that human beings are entitled to access to water regardless of their ability to pay. Any keeper of animals who failed to provide them with water would be prosecuted for causing unnecessary suffering.

Notes

1. Jean Robert, *Water is a Commons* (Mexico D. F.: Habitat International Coalition, 1994), p. 18.

2. Peter Kropotkin, *Anarchist Communism: Its Basis and Principles* (first published 1887) (London: Freedom Press, 1987), pp. 43–4.

3. Sandra Postel, *The Last Oasis: Facing Water Scarcity* (London: Worldwatch/Earthscan Publications, 1992), p. 156.

4. Comments in *Guardian*, 21 August 1995, by Frank Dobson MP on *Water Conservation: Government Action* (London: Department of the Environment, 1995).

5. Robert, *op. cit.*, pp. 23–4.

6. F. J. C. Amos, 'Planning and managing urban services: Quality standards in water supply', in Nick Devas and Carole Rokodi (eds), *Managing Fast-Growing Cities* (Harlow: Longman, 1993), p. 138.

7. *Ibid.*, p. 138.

8. Fred Pearce, *Watershed: The Water Crisis in Britain* (London: Junction Books, 1982), p. 34.

9. *We Didn't Wait for the Rain* (London: National Water Council, 1976), p. 6.

10. *The 1975–6 Drought* (London: National Water Council, 1976).

11. Pearce, *op. cit.*, p. 32.

12. Leading article in *East Anglian Daily Times*, 22 August 1995.

13. Nicholas Schoon, 'Water companies set up trust to play Santa to the poor', *Independent*, 24 December 1995.

14. *Guardian*, 12 September 1992.

15. *Guardian*, 9 November 1994.

16. Nicholas Bannister, 'Anglian Water in big deal in Buenos', *Guardian*, 13 January 1993.

17. *Guardian*, 12 September 1992.

18. Bannister, *op. cit.*

19. Erik A. Swyngedouw, 'The contradiction of urban water provision: A study of Guayaquil, Ecuador', *Third World Planning Review*, Vol. 17, No. 4 (November, 1995), pp. 387–405.

20. *Ibid.*

21. *Ibid.*

22. *Ibid.*

23. *Ibid.*

24. *Ibid.*

25. David Lehmann, *Democracy and Development in Latin America* (Cambridge: Polity Press, 1990).

26. Leonard Bays (Secretary General of the International Water Supply Organisation) in Mary Monro (ed.), *Water Technology International* (London: Twentieth Century Press, 1991), pp. 11–12.

27. Héctor Woollands, *Reseña de Actividades de la Comisión Vecinal Pro-Acueducto Sur* (Lausanne: CIRA, 1990).

28. *Ibid.*

29. M. Zolezzi and J. Calderon, *Vivienda Popular: Autoconstrucción y Lucha por el Agua* (Lima: DESCO, 1985).

30. Paul Herrington, *Pricing of Water Services* (Paris: OECD, 1987).

31. *Ibid.*, p. 9.

32. 'Water meters, secret agenda', *Guardian*, 21 February 1996.

33. Herrington, *op. cit.*, p. 21.

34. *Ibid.*, p. 17.

35. *Ibid.*, p. 19.

36. *Ibid.*, p. 23.

37. *Ibid.*, p. 21.

38. 'Water companies face challenge on meters with automatic cut-off', *Guardian*, 5 April 1996.

CHAPTER 9

The Unequal World of Water

> *Paul had turned off the freeway now, and as he drew nearer to his goal the houses grew larger, the lawns wider and greener, and the underground sprinklers rose into higher fountains, as if heralding his coming. He had noticed this before, driving past the dry barren yards of the slums near the Nutting Corporation, and down in Venice Beach, where the taps in the kitchen often gave only a brownish trickle, and no one could afford to water anything larger than a potted plant.*
>
> **Alison Lurie, *The Nowhere City*[1]**

Alison Lurie's novel *The Nowhere City*, from which this chapter's opening quotation is taken, is concerned with an American migrant from the East Coast who actually notices, when he moves to California, that unlimited water is available for those who can afford to pay for it. Most travellers tend to take it for granted that, wherever they are in the world, there will be water from the tap. They think about public services only when they are absent or restricted.

A much-travelled friend of mine, Ken Worple, keenly observes the water services that sustain the places he visits. Cycling in the hot summer of 1990 across the vast plains of Hungary, he found that 'happily all Hungarian villages contain large numbers of bright-blue water hydrants in the street which can be pumped to provide eruptions of cold drinking water, which can also be used to dowse one's head, body and feet'. Flying into Alice Springs, in far wealthier and more sophisticated Australia, he saw 'suburban houses that soak up more water per day than is piped to Aboriginal out-stations in a year'.²

Few of us lament the passing of the former Hungarian regime, but its rural water policy exemplifies the Oxfam principle that 'The aim should be *some for all, rather than all for some*.'³ I have yet to meet a water provider who is prepared to defend the market criteria that determine the distribution of water in Alice Springs, or California, Latin America or South Africa. But today's advocates, aware of the conflicts between equity and economic efficiency, have to provide an economic argument for minimal public provision:

> *For example, provision of public standpipes which provide cheap (or free) water to low-income groups could increase economic efficiency (by reducing the time wasted in collecting water and reducing health losses caused by consumption of polluted water) as well as improving the position of the poor and hence achieving greater equity.*⁴

Despite the economic case for social justice in access to water and sanitation, the situation of the urban poor has steadily worsened. For the authors of this comment also report that out of 58 developing countries in a survey conducted by the UN Centre for Human Settlements in 1986, 26 showed a lower proportion of the urban population with access to clean water in 1980 than in 1970. 'The World Health Organization estimated in 1985 that 25 per cent of the population of Third World towns and cities still lacked access to safe water, meaning that 100 million more people were unserved in 1985 than in 1975.'⁵ There is reason to believe that the situation was worse than the statistic revealed, and that it has deteriorated since then. David Satterthwaite, director of the Human Settlements Programme of the International Institute for Environment and Development, questions the basis of official assumptions:

> *Official United Nations statistics suggest that by the early 1990s more than 80 per cent of the urban population in Africa, Asia and Latin America were 'adequately served' with piped water. But these statistics considerably overstate the number for two reasons. The first is the lack of an agreed definition of what is 'adequate' and the latitude given to governments in making the judgment. The second is*

the tendency for governments greatly to exaggerate the proportion of their people with piped water supplies. United Nations agencies such as the WHO, being inter-governmental, are obliged to publish the water supply statistics supplied by their member governments. Staff from the WHO know better than anyone else the inaccuracies in the statistics they publish, but they cannot publicly question them.[6]

Satterthwaite shows that what applies to water supply is equally true of sanitation, and consequently, too, of the health burden faced by low-income urban dwellers. And he reminds us that community involvement in these services is nowadays prescribed for the poor, to reduce costs, but not for the affluent: thus,

both government and international agencies often support the installation in low-income neighbourhoods of water, sanitation and drainage systems that the households or the wider community has to manage and maintain. But there has been no trend towards asking middle- and upper-income groups to help manage their water supply networks, sewers and drains.[7]

Nevertheless, he puts his faith in small, local and unofficial initiatives, since

many of the innovative projects that reached low-income groups with improved housing and basic services came from local non-governmental organisations or municipal authorities in the South, many implemented with no support from aid agencies . . . The enormous failure to provide water, sanitation and health care and the huge health burden these impose on urban populations seem unmanageable when aggregated. But if these problems in each city and municipality can be addressed by more competent and accountable local authorities working with citizen groups, NGOs and other local participants, the problem appears more manageable.[8]

The example from Mar del Plata that I cited in Chapter 8 illustrates the importance of local citizen action in hastening the machinery of water provision and in demonstrating that local popular control *can* be more effective and less dependent on the international money market than reliance on the official machinery. But throughout the world, current wisdom demands that both city and country should balance governmental budgets by trading in the world market. This results in rural areas in a concentration on cash crops for export rather than on subsistence crops for feeding the family. In countries with an arid climate the needs of poor peasants usually take second place to the demands of irrigated agriculture and horticulture.

The same thing is true of the impact of tourism. Governments seek to open up new markets for the global tourist trade for the sake of the foreign currency it brings. Landowners and entrepreneurs of every kind are equally aware that catering for visitors from the rich world is more lucrative than sharing the products of local farming within the local market. Thus, in his report on the environmental impact of tourism, Jonathan Croall reveals that

> *Thousands of peasant farming families have been turned off their land to make way for the increasing number of golf courses being built in south-east Asia to cater for tourists and others. As a result, Amnesty International has had to create a new category of 'golfing' prisoners of conscience. In Thailand between 1989 and 1994, local developers created 160 courses, most of them out of rice fields in agricultural regions. Some are eighty times the size of a typical European course, and include luxury hotels, conference centres, fitness suites and even private airstrips. Though the golf courses are supposed to be self-sufficient, it is claimed that they take water from reservoirs illegally, or dam up streams that flow through them towards the reservoir, so making present drought conditions worse.*[9]

Golf is, ironically, the measure of success in one of the key issues of water provision in waterless places. Desalination of sea water, which is 90 per cent of the water in the world, is the apparently obvious answer to the demands we make on water supplies. The problem is the simple one that all available techniques for meeting human needs this way are frighteningly expensive. There are communities in the Mediterranean, like Gibraltar, where it has to be used, in conjunction with sophisticated harvesting of rainwater, and it is extensively used in the oil-rich Gulf States, where the national income can pay for desalination plants and their high running costs. The 'European Golf Tour' visits the best courses around the globe and, after North Africa, shifts to the 'highly rated facilities' of the Emirates Golf Club in Dubai before going on to Thailand. The Dubai golf course consumes half a million gallons (227,500 litres) of desalinated water a day.[10]

In the same water-short region, the pioneer Jewish re-settlers in Palestine sought to bring sophisticated agriculture to a neglected corner of the former Ottoman empire, to make the desert bloom and to share their experience with their Arab neighbours.[11] In the first of these endeavours they succeeded remarkably well. Modern Israel is the country where we have to learn how to make the most of accessible water, and how to cope with the problem of mineral salts in heavily irrigated land. Every visitor is made familiar with triumphs of irrigation. Howard Jacobson reported that 'I have been here only a couple of days but already a dozen different people have talked to me about the delayed-timing sprinkler-system perfected by Israelis and the envy of a water-needy world.'[12]

But the familiar intrusion of politics and nationalism has distorted the concept of fair shares of a common good. Thus Abdul Rahman Tamini, director of the Palestine Hydrology Group, makes an observation on the paradox familiar all over the world, of grossly unequal rights to water: 'You won't find any Israeli settlements without water – they have lush lawns, even swimming pools – while there are scores of Arab villages with wholly inadequate supplies.'[13]

The implications of this visible and public inequality are explored in Chapter 5, but the issues go further than lawns and swimming pools. Policies of international bodies like the World Bank and the International Monetary Fund as well as the General Agreement on Tariffs and Trade (GATT) have determined that the future of the 'developing' countries depends upon the export of primary products to the rich nations, for the sake of the debts accumulated by their ruling élites. Many years ago, when the development of by-products of the petrochemical industry destroyed the world market price of sisal, Julius Nyerere, then president of Tanzania, asked the question, 'What are we to do with our sisal? We can't eat it.'

This is the issue that arises in country after country as international bodies have demanded cash crops for export, rather than subsistence crops for local consumption. As Joan Davidson explained,

> The ecological costs of the headlong rush to generate more foreign exchange have been exacerbated by deteriorating terms of trade. Take the case of cocoa, one of West Africa's main sources of foreign exchange. Between 1985 and 1989, the region increased cocoa production by 26 per cent, as governments diverted resources to the export sector. But over the same period, the equivalent of more than US$3 billion was lost in foreign exchange as world prices for cocoa fell to their lowest-ever levels in real terms. Exporters of coffee and cotton have also seen falling prices wipe out the gains they anticipated from increased production . . . Every year, the pressure is on to produce more and more, yet this earns less and less.[14]

The disasters that follow centrally directed concentration on cash crops for export do not end with financial disappointment. Everywhere in the world peasant farmers have relied on growing staple foods for local consumption and have adapted their practices to the available water supply and the availability of fertilizers. But the major export crops not only demand much more in nutrients – which results in expensive imported agro-chemicals, which in turn can lead to groundwater pollution – but also require much more water than local food crops. The rich world's demand for cotton has for centuries created human disasters. Thus the editors of *The Ecologist* concluded that

> The repercussions of the cotton trade were catastrophic and affected people of almost every hue and clime. In the US, about 90,000 Cherokee Indians were evicted from their lands to make way for cotton plantations, 30,000 of them dying on the march west. The period 1784–1860 saw an eightfold increase in the number of slaves in the Southern states, specifically for the cotton plantations, an increase which came to a climax in the most bloody conflict of the 19th century.[15]

The process is not merely a matter of history. We have seen how it was the desire to produce more cotton that lay behind the series of attempts to harness the Nile, which are bringing endless political and social problems to Egypt and its neighbours, just as it destroyed the Aral Sea as a result of the policies of the former Soviet Union. And *The Ecologist* brings the story up to date:

> Under Ethiopia's Third Five Year Plan, 60 per cent of the lands brought under cultivation in the fertile Awash Valley were devoted to cotton production. The local Afar pastoralists were evicted from their traditional pastures and pushed into fragile uplands, contributing to the deforestation that has been partly responsible for Ethiopia's ecological crisis.[16]

The fortunes that the cotton industry has amassed for centuries have never, in any continent, benefited the men, women and children, employed to plant, nurture and harvest it, and whose water supply is diverted to nourish it. Subsistence farmers and herders are driven out to make room for the sake of an overloaded world market. Joan Davidson made this clear from the experience of other African countries like Burkina Faso, Chad and Mali:

> In the Koutiala area of Mali, problems of environmental degradation have been accelerated by the development of cotton production which has led to the clearance of wooded areas and encroachment on land previously used for food crops. The result of this pressure on resources is a shortage of land for small-scale farmers and a breakdown of the traditional system of agriculture which allowed fields to rest by remaining fallow. Because of land clearance, there are few trees left for fuel and the villagers have to use cow dung and cotton stalks, which would otherwise be used to replenish soil fertility in an area already vulnerable to drought and soil erosion.[17]

Old-fashioned imperialism is dead, but has been replaced by a far more efficient economic imperialism, which obliges the poor world to destroy its precarious economy and environment, to benefit the consumer economy of the rich world. Water which could be managed to provide a local livelihood is squandered for the sake of a highly competitive export market or for the tourist industry. And, as always, the casualties of the global market economy are the local populations, deprived of a water supply for the exclusive benefit of strangers. It is left to unofficial charitable organizations like Oxfam to insist that the elementary principle of using water for human needs is that of some for all, rather than all for some.

Notes

1. Alison Lurie, *The Nowhere City* (London: Heinemann, 1965), p. 45.

2. Ken Worpole, *Staying Close to the River* (London: Lawrence & Wishart, 1995), pp. 99, 139.

3. Jan Davis and Gerry Garvey, *Developing and Managing Community Water Supplies* (Oxford: Oxfam, 1993), p. 29.

4. Nick Devas and Carole Rokodi (eds), *Managing Fast-Growing Cities* (Harlow: Longman, 1993), p. 61.

5. *Ibid.*, p. 49 citing *Improving environmental health conditions in low-income settlements: A community-based approach to identifying needs and priorities* (Geneva: WHO, 1987).

6. David Satterthwaite, 'The underestimation of urban poverty and of its health consequences', *Third World Planning Review*, Vol. 17, No. 4 (November, 1995), pp. iii–xii

7. *Ibid.*, p. ix.

8. *Ibid.*, p. x.

9. Jonathan Croall, *Preserve or Destroy: Tourism and the Environment* (London: Calouste Gulbenkian Foundation, 1995), p. 46.

10. David Davies, 'Putting up a storm in the desert', *Guardian*, 28 January 1994, and BBC *Today* programme, 27 January 1994.

11. See, for example, the early literature such as Henrik Infield, *Cooperative Living in Palestine* (London: Kegan Paul, 1946), or Maurice Pearlman, *Collective Adventure* (London: Heinemann, 1938).

12. Howard Jacobson, *Roots Schmoots* (London: Viking/Penguin, 1993), p. 283.

13. *Guardian*, 31 May 1991.

14. Joan Davidson and Dorothy Myers, *No Time to Waste: Poverty and the Global Environment* (Oxford: Oxfam, 1992), p. 55

15. The Ecologist, *Whose Common Future? Reclaiming the Commons* (London: Earthscan Publications, 1993), p. 136.

16. *Ibid.*, p. 136.

17. Davidson and Myers, *op. cit.*, p. 167, citing W. Critchley, *Looking After Our Land* (Oxford: Oxfam, 1991).

CHAPTER 10

Dirty Water

> *It may be just another statistic, but did you know that Britain alone pumps over 300 million gallons of sewage into the sea around our coasts every single day? Hardly surprising that people now think twice before taking a dip. And at what cost do we flush? Water companies spend millions on purifying water for us to use and much of that beautifully clean water goes into flushing the indoor toilet. If that is not a criminal waste, then I am indeed more ignorant than I thought I was.*
>
> **Jean Turner,** ***East Anglian Privies***[1]

I live in rural East Anglia, where everyone over the age of 50 remembers the privy, or earth closet, a shed in the garden with a seat and a bucket of earth. Town-dwellers and the rich might have moved long ago into the era of the water-closet, but villagers and isolated rural dwellers reached it long after they had the blessing of a piped water supply. When Jean Turner wrote to the Ipswich and Norwich papers and the local federation of Women's Institutes for recollections of the days of the privy, she was deluged with correspondence.

The stories she gathered fall into a pattern, with plenty of comments on the discomforts, and about the ingenious devices for using ashes from the stove to cover the offensive faeces, as well as about the marvellous vegetables their parents produced from the careful recycling of human wastes, as people around the world have learned to do from the dawn of horticulture. She kept her own telling conclusion, which I quote at the head of this chapter, for her very last page, together with a commendation for the people who are exploring ways of introducing 'biological closets' to convert faeces into 'a productive, economic, environmentally friendly and safe compost'.[2]

In Britain this is still seen as a marginal issue, but it will certainly grow in importance in the twenty-first century. Inland cities, because of the threat of epidemic disease, developed elaborate systems of sewage processing and sought, regardless of the inhabitants' ability to pay, to link every urban household with the sewage system. Coastal and estuarial towns frequently discharged untreated sewage into the seas around the British Isles, because it was the cheapest way of coping with the problem, and with the assumption that the seas were big enough to absorb and dilute everything from untreated faeces and industrial wastes to the effluents of the nuclear industry.

It used to be said with pride that the safe water consumed by Londoners had been through the supply and disposal systems of several upstream towns on the Thames before it reached their homes. One legacy of the triumphs of Victorian sanitary engineering is that water charges are much lower for the inhabitants of inland cities like Birmingham than for those of counties with a long sea coast like Devon and Cornwall. When they were sparsely populated, nobody worried about the disposal of wastes into the sea. Now that their much-increased population is swollen by the holiday trade every summer, swimmers, surfers and families on beaches are outraged by the visual evidence of raw sewage when they explore the shore or the sea.

For many decades the British took for granted their water supply and drainage systems. They were provided by the expertise of municipal engineers and gradually extended to meet the needs of rural areas, and were a very small charge on local taxation. Water engineers assumed a continual growth in demand, and planned reservoirs and extraction plants to meet future needs. Some municipal engineers held to the view that the natural and historic use of sewage sludge was as a fertilizer, but market considerations suggested that the cost of pulverizing and drying the material, as well as high freight charges, would be higher than any revenue resulting from the sale of processed sludge.

The 1970 government report cited in Chapter 1 found that, beyond the fact that in some areas sludge might be excessively contaminated by toxic metals,

> Farmers usually prefer to use artificial fertilisers since they are not only of constant composition and consistency, being in a form suitable for assimilation by the soil, but they are completely balanced in respect of the essential nutrients for the particular crop. They also attract a high subsidy payment . . . Sewage sludge is deficient in one of the three essential nutrients, namely potash. It has been suggested that if this deficiency in potash were remedied, then not only would the product be more acceptable to the farmer, but it would attract a reasonable subsidy. This we find is not so, since no subsidy is payable on potash . . . We conclude that the application of sewage sludge to agricultural land is of restricted use. But where it is practicable, it is a good method of disposal and one which should be encouraged, utilising the limited fertilising properties of the sludge. We suggest that local authorities should investigate and embark on more positive marketing methods . . .[3]

A quarter of a century later the scene has changed. Not only have the water supply and sewerage industries been sold in the private market, but we also have an environmental lobby, concerned with economy in water supply and ecological issues in sewage disposal. We have all become conscious of the absurdity that 32 per cent of domestic consumption is the use of an expensively purified product for flushing water-closets.[4]

Another change results from the British government's undertaking to conform to standards required by international agreements. The European Union, in its former capacity as the European Community, has issued a long series of water standards whose great value has been in providing ammunition for the unofficial campaigning organizations. For example, it issued directives on the quality of bathing water at the seaside, based on physical, chemical and microbiological quality. The Marine Conservation Society and the Coastal Anti-Pollution League issue their *Good Beach Guide*, and the Tidy Britain Group administers the Blue Flag scheme. Efforts by the British government to dilute the effectiveness of these assessments are described by a leading authority on water quality, David Kinnersley, as 'blatant fudging'.[5]

Similarly, the importance of the European directives on pollution control of rivers and on the quality of drinking water have had a similarly salutary effect. Mr Kinnersley remarks on the irony that, just as tap-water is subject to more thorough independent monitoring than ever before, 'the British have taken to drinking bottled water on a scale hardly imagined earlier. Moreover, they do so at supermarket prices which, litre for litre, costs more than 1,000 times the price of tap-water.'[6]

It was similarly reported in 1996 that hotels and restaurants were selling bottles of water produced by connecting a filter system to the ordinary water supply at a profit of more than a thousand per cent.[7] These are the follies of a rich society, where there is little interest in separating the water used for drinking from that used for water-borne sanitation, or car washing. Separate pipe systems are unlikely, but much could be saved by a reduction in the quantity of water used, or a smaller flush for urine than for faeces.

Non-waterborne sanitation systems do exist in the rich world, mostly in the form of individual composting reactors.[8] Some authorities on sustainable patterns of settlement see a great future for the Memtech system of local composting pioneered in Australia.[9] But few are optimistic enough to see these sophisticated systems becoming available to the vast cities of the poor, where the need is most urgent.

I mentioned in Chapter 5 the work of the architect Jean Robert, whose activities for years have been centred around local promotion of safe non-waterborne methods of sewage disposal. He reminds us that

> *Much of the skyrocketing debt of the poor countries is fueled by technologies whose only aim is to understand what the flush toilet does, namely mix water with shit. Meanwhile, the inhabitants of great Latin American cities like Lima continue to live above an underground sea of black water.*[10]

Should we be surprised that half a million cases of cholera were reported in Latin America in 1991, seventeen thousand of them in Peru? In Mexico City a tunnel was bored through the mountains to take away rainwater to the Mezquital valley, growing the vegetables sold in the city's markets. 'By now, however, the canal has become a huge sewer draining a mixture of untreated city excrement, heavy metals and detergents down to the valley where this mixture irrigates gardens and fields.'[11] So in 1992, for fear of the incipient cholera epidemic, the government forbade the sale of the valley's vegetables, threatening the very existence of the cultivators.

For Jean Robert the spread of cholera is merely the tip of the iceberg in the progressive degradation of health conditions in urban areas. He stresses that the most common water-related disease is dehydration, but that

> *In a world inhabited by some five and a half billion people, more than one third do not have safe drinking water and a quarter have no form of sanitation. Some 50,000 deaths occur every day from waterborne diseases. To put it in perspective, that's a third of all deaths occurring in the world. Forty per cent of Africans are in danger of contracting water-related infections, with the main causes being the presence of excrements and toxic chemicals in their water supplies . . .*[12]

He argues that when indoor plumbing was a luxury available only to the rich, it was found cheap and convenient to discharge faeces into the rainwater drainage system, and that this was the first step in the 'globally insolvable problems' that we face today. The last was that of adding industrial wastes to this dangerous mixture. 'The compound of biologically highly active organic matter and toxic chemicals thus obtained is an epidemiological and ecological time bomb that repair technologies are manifestly unable to de-activate.'[13]

Whether effective technology is available or not, the political will is absent. A glance at the press any week indicates that politicians are unwilling to enforce existing legislation. Thus in 1995 an organization in the United States called Project Relief sought to forbid the use of public funds to enforce existing legislation to set water quality guidelines, limit the dumping of sewage in rivers, or to control air pollution or regulate the amount of pesticides permitted in foodstuffs. Martin Walker reported that 'A coalition of 115 corporate groups and industrial lobby groups, Project Relief gave $10.3 million to Republican congressional campaigns.'[14]

The British situation is less blatant, but there is an obvious unwillingness to undertake 'repair technologies'. The water industry regulatory body, Ofwat, found in November 1995 that 'nearly a quarter of the drinking water in England and Wales still fails to meet pesticide standards'.[15] Similarly, a long series of reports suggest that while increased charges for water are explained by the need to reduce pollution, the work is not being done. Thus in 1993 it was alleged that 'Water companies are cheating consumers by forcing them to clean up pesticide pollution when farmers and the chemical industry are responsible.'[16] And two years later, leaked government documents revealed that 'water treatment works covering pollution at bathing beaches and in rivers have been delayed despite increased charges, ostensibly to pay for them'.[17]

The European requirement on acceptable quality on beaches was avoided by the British government's claim that the definition of a beach

> *should be applied only to places where 500 bathers were actually in the water at any one time or where there were more than 1500 bathers per linear mile of beach. This definition excluded not only all Welsh beaches from the category of beach but also excluded Blackpool – the United Kingdom's best-known seaside resort.*[18]

A further requirement of the European Union, accepted by the British government in 1991, set standards for the discharge of urban waste water into rivers. A loophole in the directive provided for less stringent standards for 'high natural dispersion areas' where the sea would quickly carry away waste. In 1994 the Secretary of State for the Environment, John Gummer, performed 'a bizarre geographical shuffle to enable the privatised Yorkshire Water Company to escape an obligation to build a £100m new sewage works to clean up [Hull's] pollution'.[19] He declared more than 30 miles (48km) of the River Humber to be open sea, so that it could continue to receive crude sewage from the city of Hull. He repeated his conjuring trick by making a similar ruling for Bristol on the River Severn. Both city councils objected, and in 1996 the High Court ruled that this attempt to evade environmental legislation was 'unlawful'.[20]

This governmental manipulation of geography is precisely the same response to attempts to control the discharge of human and industrial waste products that impels manufacturing industries to relocate plants from the rich nations with environmental legislation to poor nations for whom such controls are either lacking or are unenforceable. In this sense Jean Robert was right to claim that we are manifestly unable to deactivate the epidemiological and ecological time bomb. Political leaders will not accept the consequences of the legislation they have been obliged to enact.

For just as the suppliers of water are unwilling to entertain the cost of the installations and pipework to provide water of an expensively purified quality for culinary and drinking purposes and a separate system of less-treated water for the other uses that form the greater part of domestic and industrial consumption, so the disposers of water-borne waste products are unwilling to meet the costs of separating human and industrial effluents. In Britain this was made clear by the Chairman of the Royal Commission on Environmental Pollution. In introducing its report on the sustainable use of soil, he observed that

> *The Commission is concerned about the effects of sewage sludge contaminated with chemicals and heavy metals being spread on the land as fertiliser but potentially poisoning it. This arises partly out of the Government's policy of allowing industry to mix its discharges with domestic sewage, a policy the Commission wants stopped.*[21]

However, the Secretary of State for the Environment had already announced that the Royal Commission's recommendations for 'large-scale schemes to clean up rivers and protect public water supplies from industrial pollution will not be implemented', but the task would be left to the long-term programme of the National Rivers Authority.[22] Then in April 1996 the National Rivers Authority was subsumed, together with Her Majesty's Inspectorate of Pollution, and various waste regulation authorities, into a new body, the Environmental Agency. Environmental campaigners see this measure as an attempt to insert political 'realism' into the pollution debate and to apply 'the techniques of cost-benefit analysis to its recommendations'.[23]

Governments committed to promoting untrammelled market forces at any social cost are unlikely to heed the recommendations of bodies like the Royal Commission on Environmental Pollution, unless the sheer weight of public outrage obliges them to do so.

Notes

1. Jean Turner, *East Anglian Privies* (Newbury: Countryside Books, 1995), p. 125.

2. *Ibid.*, p. 125.

3. *Taken for Granted*, Report of the Working Party on Sewage Disposal (London: HMSO, 1970), p. 68.

4. David Kinnersley, *Coming Clean: The Politics of Water and the Environment* (Harmondsworth: Penguin Books, 1994, p. 94.

5. *Ibid.*, p. 129.

6. *Ibid.*, p. 160.

7. Ray Clancy, 'Profits flow from filtered tap water', *Daily Telegraph*, 7 February 1996.

8. See, for example, Edward Harland, *Eco-Renovation: The Ecological Home Improvement Guide* (Devon: Green Books, 1993), p. 135.

9. Herbert Girardet, *The Gaia Atlas of Cities* (London: Gaia Books, 1992), p. 164.

10. Jean Robert, *Water is a Commons* (Mexico D. F.: Habitat International Coalition, 1994), p. 36.

11. *Ibid.*, p. 38–9.

12. *Ibid.*, p. 35.

13. *Ibid.*, p. 27.

14. Martin Walker, 'Licence to pollute the free world', *Guardian*, 6 September 1995.

15. Chris Barrie, 'Quarter of tap-water fails to pass pesticide tests', *Guardian*, 25 November 1995.

16. James Erlichman, 'Water clean-up cheats public', *Guardian*, 15 November 1993.

17. Nick Nuttall, 'Water firms are cheating customers', *The Times*, 13 November 1995.

18. Robin Clarke, *Water: The International Crisis* (London: Earthscan Publications, 1991), p. 26.

19. Geoffrey Lean, 'John Gummer turns Hull into a seaside resort', *Independent*, 13 November 1994.

20. *Guardian*, 27 January 1996.

21. Sir John Houghton, Chairman of the Royal Commission on Environmental Pollution, 29 February 1996, introducing the Commission's report on *Sustainable Use of Soil* (London: HMSO, 1996).

22. Paul Brown, 'Schemes to clean up rivers and protect drinking water blocked', *Guardian*, 24 February 1995.

23. Christian Wolmar, 'Ministers axe plan to cut pollution', *Independent*, 2 April 1996.

CHAPTER 11

A Confluence of Crises

There is enough for everyone's need, but not for everyone's greed.

M. K. Gandhi, *The Constructive Programme*[1]

Some of us are automatic optimists, believing that human ingenuity can find a solution for every problem. Others are natural pessimists, confident that human folly makes it inevitable that all our problems will get worse. Sir John Houghton is neither. He is professor of atmospheric physics at Oxford, a former director of the Meteorological Office, chairman of the British government's Royal Commission on Environmental Pollution, and of the United Nations advisory panel on climate change.

In February 1996 he addressed the Royal Society in London on the impact of global warming, a topic of endless speculation. He explained that his panel had concluded that since the 1970s there had been a relatively steady rise in global temperatures and that this was probably the result of a build-up of heat-trapping pollutant gases in the atmosphere. This process was likely to increase in the twenty-first century, and the expansion of water in the oceans would raise sea levels by about two feet by the year 2100. He stressed that while it was still difficult to predict the results in detail, all projections indicated increased rainfall in the regions which already have monsoon conditions, and more severe drought in those already prone to water shortage. He observed that 'Adaptation will be very difficult . . . in some particularly vulnerable areas such as the delta regions of large rivers in Bangladesh, Egypt and Southern China, and the many low-lying islands in the Indian and Pacific oceans.'[2] The expected changes might not threaten food supplies on a global scale, but would greatly affect specific populations:

> Some regions may be able to grow more, others less, but the distribution of production will change because of changing water availability. The regions likely to be adversely affected are those in developing countries in the sub-tropics with rapidly growing populations. In these areas there could be large numbers of environmental refugees.[3]

In other words, the people who will be the victims of global warming caused by pollutants released into the atmosphere by the rich world, will be the world's poorest inhabitants. However, it is interesting that Professor Houghton, resigned perhaps to the fact that a topic as hard to grasp as that of climatic change will be ignored by politicians until its results are upon us, also raised issues that are already with us. He reminded us that we already have innumerable water crises:

> Demand for water has been rapidly increasing in nearly every country and especially in those where it is extensively used for irrigation. There are already significant tensions, especially in regions where the water from major river systems is shared between nations. It is not surprising that the Secretary General of the UN has suggested that wars in the future are likely to be about water rather than oil.[4]

Plenty of people will decline to recognize the facts of global warming until they are personally affected by its consequences, but these are likely to be a worsening of water situations that already exist. On the worsening of international conflict, I cannot improve on the diagnosis provided by Bharat Dogra, discussing the Ganga-Brahmaputra-Barak river basins:

> Our rivers, more than any other part of our ecosystems, know no borders, yet water has increasingly been treated territorially. The cutting up of river systems by state boundaries has aggravated the problems of responding ecologically to floods. Instead, water conflicts are turned into political conflicts between states. However, between people that share a river, there is no conflict. Fisherfolk and peasants living on the banks of the Ganges in India and the poor peasants

> living on its banks in Bangladesh are connected to each other through the life of the river. The political and engineering structures that threaten their lives and livelihoods, simultaneously bring economic power and political control to national and international elites. Years ago, when no line divided Bengal, and no engineering structures had cut across the Ganges, Tagore had written a play called Mukta Dhara (The Liberated River), *in which he had symbolised colonial rule through the dam, and Gandhi's struggle for freedom from dependency and control as the liberation of the river.*
>
> The metaphor is now more relevant than ever. Unless the people of the Ganges basin are able to co-operate with one another and seek a solution to floods which is in harmony with the ecosystem, we will have to passively accept costly cures that will at some point become even worse than the disease.[5]

His account describes precisely the situation of the people most threatened by global warming, imposed upon them by the profligacy of the rich world.

The irrigation issue is equally complex, and reveals the same conflict of ideologies. There are communities whose agricultural economy has been constructed around the management of geology and seasonal rainfall to produce water-demanding crops, and there are communities who have had to follow the demands of the market economy, neglecting production of local subsistence for the sake of cash crops for export. All over the world the point of diminishing returns has been reached as mineral salts steadily increase the salinity of the vital water and reduce output.

Some authorities see an increasing reliance on irrigation. Others predict that the peak has already been passed. For example, in the *Dictionary of Environment and Development*, Andy Crump reports that

> By 1990, 30–40 per cent of the world's irrigated cropland was thought to be either waterlogged or suffering from excessive salinization . . . Although water withdrawals continue to increase, by 1991 a decreasing proportion of withdrawals were used for irrigation. The 63 per cent of water being used in 1991 was projected to decline further: to 55 per cent by the year 2000.[6]

His conclusion was that

> *it is generally recognized that irrigation projects are now most likely to succeed where fallow periods are observed and management is left up to local communities — irrigation schemes should be small- rather than large-scale.*[7]

The message is true, no doubt, but in the short-term world of global food markets where traders profit and producers go hungry, it is slow to be absorbed. Reliance on large-scale irrigation has spread from luxury export crops in dry climates to the production of ordinary vegetables for the supermarkets that handle most of the retail trade in Britain, a country which by world standards has ample rainfall. Large-scale farmers are encouraged both by the National Rivers Authority and the Ministry of Agriculture to build their own reservoirs and to abstract water from rivers with a 90 per cent reduction in the charge made if they take it in winter rather than in summer.

This is precisely the foresight and wisdom advocated for the poor world, but its effect in the rich countries is simply to drive small growers out of the market. Thus in the dry summer of 1985 it was reported that

> *The eight-week life of a supermarket lettuce involves lavish supplies of water. Every plant is drenched in three-quarters of an inch (2 cm) every seven days. Baking potatoes are even more demanding, drinking 10 inches (25 cm) of water over the summer to deliver the right size tuber free from blemishes to satisfy the specifications of the retail chains. Nick Darby, whose 1,200-acre Norfolk farm supplies iceberg lettuces to McDonald's and other varieties to almost every supermarket chain in the country, said . . . 'More reservoirs have been built in the past five years than ever before. The operation is geared to supply supermarkets, with the contracts specifying both volume and quantity. Any variation on that costs us dearly.'*[8]

It is thus more than a little sanctimonious for the rich world, eating prosaic vegetables like potatoes, parsnips and carrots, all produced through irrigation, to criticize growers in the poor world for their profligacy in irrigating export luxury crops.

The collapse of traditional heavy industries in Britain has

considerably reduced the industrial demand for water, or, more accurately, has transferred it to other parts of the globe. It is increasingly the domestic consumer who is seen as the water-waster, precisely the user we are desperately hoping to help in the poor countries of the south. A report from the National Rivers Authority in 1995 urged that, to conserve supplies,

> Water companies should give people grants to buy water-efficient washing machines and dish-washers as well as replacement toilets which flush with less water. Spending £300 per family buying each a new lower-flush toilet would be more cost-effective than building a new reservoir, while supplying the entire country with new toilets could save 13 per cent of the nation's water needs. It was no surprise that the authority found fault with the water company's failure to stop leakages, which account for 34 per cent of water wastage.[9]

Of all our water crises, the worst by far is not our profligate use of water in the rich world, but our unwillingness to deal adequately with the effluents we produce. We saw in the last chapter that the governments of the wealthiest of nations have consistently evaded this moral imperative. We are in no position to blame those of the other nations with a rich élite and a poverty-stricken population who fail to enforce intergovernmental decisions about environmental pollution.

The global problem of water is less a matter of our misuse of a natural and renewable resource than one of our unwillingness to share our wisdom with other people who desperately need safe water and release from endless daily labour in gathering it. When we have learned to see ourselves as members of a water community, rather than as customers paying the market price, we will be able to give help based on experience.

Edward Luttwak, Senior Fellow at the Centre for Strategic and International Studies in Washington, points to the ironic consequences of the fact that while the earth is abundantly provided with water, there is a shortage of available fresh water with a 'finite global supply of 9,000 to 14,000 cubic kilometres per year, versus the Earth's total stock of 1.4 billion cubic kilometres of mostly unusable water', and is at pains to stress that, for the rich world, water choices 'are much more a matter of values and customs and politics than of physiology or ecology'. To illustrate this conclusion, he stresses individual choices:

> *Choose a kilo of cotton clothes and you are choosing to use 29,000 kilograms of water to grow it (Oklahoma data – Egyptian cotton drinks more); choose nylon, made from adipec acid extracted from corn-cobs or oat hulls, and not a drop of water is needed beyond what is used anyway to grow those crops for food; choose rayon, derived from wood pulp, and you are choosing to cut down trees . . . Again, human choices, not natural limits, dominate; it takes much more water (nineteen times more, by one estimate) to feed on beef rather than wheat, though water-use in agriculture is so profligate (swamp crops such as cotton and rice are now routinely farmed in deserts by river-devouring irrigation) that steak-lovers can continue to eat steak without environmentalist guilt feelings.*[10]

Most of us are ill-equipped to enter into this kind of paradoxical calculation, and in any case resent the downgrading of public and social policy to a matter of personal consumer choices. The British experience is relevant here. When water supplies were municipally controlled or provided by private companies as 'statutory undertakers' operating without profit, it was dominated by engineering considerations. If anticipated demand was unlikely to be met, another valley was drowned to provide a new reservoir with only local opposition. The ideology was to meet future needs at any cost. When a Conservative government replaced the tapestry of local undertakings with regional water authorities, central government controlled their investment programme. When another Conservative government sold the water industry to private investors, it was seen as profligate to provide for all possible needs. This was precisely the same ideology that eliminated the warehousing of components in industry, or the provision of empty beds in hospitals.

After the drought in 1995, the Department of the Environment commissioned an independent report on the failure of water supplies in Yorkshire, asking for an assessment of requirements up to the year 2021, both with and without the effects of climate change. It found that the two largest areas of increase in water use were irrigated agriculture and domestic consumption. One of the independent assessors, Paul Herrington of the University of Leicester, explained that

> It is clear that there is not going to be enough water to go round. The government will have to have a series of strategies in place so that water companies can restrict demand. It seems that the expenditure that would be required on new reservoirs to allow people water on demand is just not available, even if it were possible to achieve a building programme on such a vast scale.[11]

He is raising a vital issue in the politics of water. I showed in Chapter 8 that when water supply in Britain was a public service, people responded willingly to pleas to reduce consumption in 1976, but that the climate had changed in 1995, when water had been downgraded from a public service to a commodity. Do water-users really want the flooding of more valleys in remote places to satisfy their demands, or are they resolved to become responsible consumers?

The British government's response was hilarious. In April 1996 it revealed its plans to create a free market in water. The Secretary of State for the Environment, John Gummer, announced that he proposed to allow, first of all, companies using an annual 250 megalitres to buy water from a supplier other than their local company, and to extend this eventually to every consumer. He explained that 'Competition is the best guarantee for consumers that they receive value for money, better service and lower prices.'[12] He said nothing about the desirability of limiting the demand for water. Speaking for the parliamentary opposition, Frank Dobson commented that

> These proposals have nothing to do with the fundamental failings of the privatised water industry with its soaring prices, profits and bosses' pay and perks. It does nothing to stop the scandal of the environmental damage of taking too much water from rivers and lakes during dry spells.[13]

His party, however, has announced no plans for the de-privatization of the water supply industry if and when it takes office, nor which of the previous regimes of water distribution (described in the British context in Chapter 1) it supports. There is indeed a void in organizational thinking in the field of community control of local services. The alternatives are perceived as public bureaucracy or private profit.

A week before the British parliament debated the proposal to introduce competition in the water industry, the municipal council of the town of Grenoble in south-east France was considering a proposal to return it to municipal control. France, like Britain, experienced a vogue for marketizing public services in the 1980s, with two huge companies, Compagnie Générale des Eaux and Lyonnaise des Eaux, as the beneficiaries. Grenoble's municipal water undertaking was sold to the latter company in 1989, and the experience of the citizens has been much like that of British equivalents, with the additional factor, reported in 1994, that 'Lyonnaise subsidiaries in Grenoble and Lyon are suspected by local judges of funnelling funds to conservative parties in exchange for contract favours. No illegality has yet been proved.'[14] However, there was widespread public outrage, as in Britain, at the results of privatization, and a coalition of opposition councillors called *Démocratie/Ecologie/Solidarité* tabled a motion demanding a return to public control of water. They were outflanked when the Mayor announced a deal for a mixed water economy, with the town owning 51 per cent and Lyonnaise des Eaux 49 per cent, accompanied by a retrospective reduction of water bills from last January 1st. The coalition split apart over this deal, which some of them claimed as a victory, while others saw it as a defeat.[15]

The British have been led to believe that France was the most centralized country in Europe, a dubious honour that should undoubtedly be awarded to the United Kingdom. It is impossible to conceive any British city regaining a controlling interest in its own water supply in the current political climate. But the devastating conclusion we are bound to draw from the insular debate is that the crisis of social responsibility has been reduced to a matter of buying in the cheapest market. A hundred and fifty years before the parliamentary debate on competition in the water industry, as the opening chapter of this book stressed, it was evident that 'water is as essential to the health and comfort of mankind as the air we breathe, and when mankind congregate in masses counted only by tens of thousands, it is essential to public health that it should be most abundant'.[16]

This old concept has been lost in both rich and poor countries. Decisions on the availability of clean water are made on economic, rather than social grounds, unless unofficial bodies step in to help local activists. If the price mechanism determines the allocation of water, the poor will die of thirst; if it determines the crops which are irrigated for market crops, the poor will die of starvation; if it determines the financial viability of pollution control, the poor will be poisoned; and if it determines the availability of water for personal hygiene, vast numbers of the children of the world's poor will die before reaching the age of five from prosaic illnesses like diarrhoea.

When Dr Christiaan Barnard, the pioneer of open-heart surgery, addressed a huge audience in a South American sports stadium, it was left to Ivan Illich to point to the statistically likely conclusion that a large proportion of his hearers were suffering from intestinal worms.[17]

We have a global problem regarding the use of water, a limited but endlessly renewed resource. There are huge engineering solutions which usually penalize local populations but benefit overseas and metropolitan investors. These in turn add to international disputes fought with a weaponry which is far more sophisticated than the simple techniques of water management. The message of this book is that if human communities actually achieved control of their own supply and manipulation of water, they would manage fairly and responsibly, recognizing the needs of all, as well as those of their fellow users of the same resource.

The tragedy of the world water crisis is that this is the last approach that the controllers of the world's water economy are willing to consider. Responsible water use depends not on pricing the poor out of the competitive market, but on following the elementary principle of fair shares for all, a concept that every child learns from infancy until it is driven out by the political realism that determines that might is right. If market forces in a climate of shortage were allowed to determine the price of water, the poor would die of malnutrition and disease, as they already do in many parts of the world, while the rich would pay up just because they are rich.

Notes

1. M. K. Gandhi, *The Constructive Programme* (Ahmedabad: Narajivan, 1945).

2. Sir John Houghton, 'Global warming: A scientific update', lecture to the Royal Society, London, 15 February 1996.

3. *Ibid.*

4. *Ibid.*

5. Bharat Dogra, *Floods in South Asia: A Report on Ganga-Brahmaputra Region* (New Delhi and Dhaka: South Asian People's Environmental Network, 1993), p. 3.

6. Andy Crump, *Dictionary of Environment and Development* (London: Earthscan Publications, 1992), p. 84.

7. *Ibid.*, p. 84.

8. Martin Whitfield, 'Farmers flood the land to satisfy thirsty vegetables', *Independent*, 30 May 1995.

9. Paul Brown, 'Water company grant urged for efficient toilets and washers', *Guardian*, 5 October 1995.

10. Edward N. Luttwak, 'Best of all worlds?', *Times Literary Supplement*, 29 March 1996, pp. 3–4.

11. Paul Herrington, interview in *Guardian*, 20 March 1996.

12. *Hansard* (House of Commons Official Report), 1 April 1996.

13. *Ibid.*

14. *Guardian*, 9 November 1994.

15. 'L'eau divise la majorité', *Le Dauphiné* (Grenoble), 26 March 1996.

16. Derek Fraser, *Power and Authority in the Victorian City* (Oxford: Basil Blackwell, 1979), p. 182.

17. Ivan Illich, *H_2O and the Waters of Forgetfulness* (Berkeley: Heyday Books, 1985), p. 3

CHAPTER 12

The Delights of Water

> *The sound of water is everywhere; cascading from lions' mouths, bubbling up into mirror-like pools, trickling and splashing from fountains and cisterns or sliding like a glittering snake through narrow open channels in the flagged floors of the rooms; the sound cooling and murmuring on the sweltering air with the onomatopoeic music of refreshment.*
>
> **William Corlett, *Now and Then*[1]**

All visitors to the Alhambra at Granada are enchanted by the ways in which a tiny flow of water is used to provide continual delights as it flows, impelled by gravity, through an elaborate system of fountains and channels, for sheer pleasure, before being fed into the gardens. The pleasures of the Alhambra are, like the immensely significant water tribunal in Valencia, a tribute to the Moorish period in Spanish history.

Jean Robert remarks, admiringly, that

> *Partly due to the fact that water is very limited in most of the countries touched by Islam, Koranic water laws are exemplary and, in Islamic countries, a still living source of inspiration for regenerative politics. The problem of Islamic legislators was: how to foster the cultural avoidance of scarcity in a context of stringent limitation? It was and is still today the core problem of any sound policy of water conservation.*[2]

If the Alhambra is a beautiful example of the means of extracting maximum pleasure from a limited flow of water, the *Jet d'Eau* in the lake at Geneva is at first sight the most profligate. Travellers by air from Britain to Venice (a city built on water and slowly sinking down into it) pass over the Alps, and while one snowy mountain is like any other for the aerial observer, the huge, man-made, waterspout at Geneva is easily identified, as it rises to 145 metres (476 feet).

Its origins were severely practical. Every year the torrents of spring brought an avalanche of melted ice and snow down to the city in the valley. So the sheer pressure of water was used to canalize it into Lac Leman, as a safety-valve to protect Geneva from annual flooding. Today, while it still services its original purpose, it is far more important as a symbol for the city, and is switched on or off throughout the summer, according to the prevailing winds and the comfort of the citizens.

We cannot leave the theme of water as a crisis of human management of natural resources without stopping to consider water as a source of sheer pleasure. Everyone who was a child in a country where there was water to bathe the children remembers the pleasures of the bath, and anyone who lives in a country where water is not a threat knows that the price or the rent of a home is enhanced if it is by the seaside, the lakeside or the riverside. In the contemporary world, where docks and canals have ceased to be economic assets, a dockside or a canalside apartment is equally valued. And in many of the world's cities the waterside is a favourite place for families, lovers and contemplative walkers.

Part of the folklore of the American city concerns the fire hydrants, illicitly opened to provide relief from the long, hot summer, and I learned years ago in Philadelphia that the opening of the hydrants is now authorized when the temperature rises to a certain level. For generations of New York children, as the summer draws on, the cry goes around, 'Who's got the spanner?' And when someone succeeds in opening up the fire hydrant, the children of the whole neighbourhood are drawn to the street and, after the joy of the initial wetting, devise elaborate rituals of splashing and squirting. By the 1960s the activity was legitimized when the Police Athletic League of New York City, in streets closed off for play, provided sprinkler heads to replace perforated tin cans. The equivalent in London in the first half of the century was following the water cart. In her history of the city's childhood, Anna Davin records how

> In summer hot children (the boys often naked) cooled off in the spray of the water carts which sprinkled roads to lay the dust, dancing and shrieking happily, sometimes with a mother walking indulgently by the side of the cart and carrying dry clothes and a towel.[3]

In much of the world, rainfall is seasonal and, since life depends on it, is desperately awaited. People wait desperately and prayerfully for the coming of the rain. Too little rain spells hunger, but too much means danger and appalling floods. People always remember the day the rain came. Jill Ker Conway, later a famous academic in Canada and the United States, was reared on a sheep farm in Australia. Her family was rich in the years when it rained, and ruined when the rains failed to appear. She remembers her fifth year, when an unheard-of eight inches of rain fell, and dams brimmed with water while sheep and cattle bloomed with health and nourishment:

> The transformation of the countryside was magical. As far as the eye could see wild flowers exploded into bloom. Everywhere one looked the sites of old creek beds became clear as the water gathered and drained away. Bullrushes shot up beside the watercourses, and suddenly there were waterfowl round about . . . We saw the sky reflected in water for the first time. Stranger still, the whole countryside was green, a colour we scarcely knew. Evidences of the fertility of the soil were all about us . . . Walks became adventures of a new kind because they were likely to reveal some new plant or flower not seen before, or show us why the aboriginal ovens were located where they were, close to what was once a stream or a water hole. We made a wooden raft and poled it cheerfully around the lake near the house, alighting on islands that were old sandhills, now suddenly sprouting grass.[4]

Water, or the lack of it, shaped her pleasures as well as her family's fortunes, and the same was true of the family of the novelist Georges Simenon. His grandfather, Wilhelm Brüll, was a 'dyke-master', living in a farmhouse beside the Zuid-Willemsvart Canal. Simenon remembered the pleasure of watching barges go past at eye level from the first-floor bedroom window, since all the land, as far as one could see, was below sea-level. He looked back with pride on the vast responsibilities that his grandfather carried:

> *From the water in the canal, from the way in which the water was shared out, from the way in which it was allowed to run along the feeder canals and then released over the land at appropriate seasons, depended the prosperity or ruin of everyone for leagues and leagues around. And Brüll was the master of the water, the maker of riches.*[5]

The thrill of being able to control the flow of water by building a dam across a stream or a gully in the street, even though the water always wins, is one of the universal delights of childhood. The growth of urban settlements was usually along a riverside, lakeside or seaside and, apart from its economic importance, the waterside has in many cities been a source of shared delight for the inhabitants. Modern London is ill-provided in this respect. When I was a child in the 1930s, small public efforts were made to provide access to water for citizens. A celebrated Labour politician was given the unimportant office of President of the Board of Works in 1929. Enquiring what his duties were, he found that he was in charge of the Royal Parks. So he set about allowing bathing in the Serpentine lake in Hyde Park. It became known as Lansbury's Lido, but in his memoirs of the period Douglas Goldring recalled that 'so mean was the attitude of the governing class that Lansbury had to fight tooth and nail against Tory obstruction to obtain for Londoners the right to bathe in their own Serpentine in their own park'.[6]

East of the City, vast areas of the waterside were walled off from the public in the huge complex of docks. There were small gaps around ancient waterside steps at Wapping and Shadwell, and the council used to bring in bargeloads of sand for children to play in at low tide, while boys would dive from the steps at high tide. Today, even though the water is actually cleaner, any child who falls in the river is given an anti-tetanus injection. It was widely believed that the obsolescence of the docks would provide, as the writer Sean French put it, the opportunity to reconceive the capital's riverside that comes only once in a century. He explained that,

Admittedly, the omens were not good. The north bank of the Thames that loops all the way from Chelsea to the Tower of London was already one of the most horrible developments along any great river in the world. The building of the Embankment, a major engineering project that occupied decades of the mid-19th century, was almost a deliberate attempt to insulate the river from the social life of the city. The underpasses and corporate headquarters that form the riverbank between Blackfriars and the Tower are even worse. The miles of prime riverfront land in Rotherhithe (and Wapping and the Isle of Dogs) came up for grabs at the very moment when the whole idea of planning for London as a whole was in disarray. It was all flogged off quickly and the result is, one can safely say, not an area that is going to be a focus of streetlife, and any attempts at streetlife will be firmly asked to move on . . .[7]

But he had a dream, inspired by a visit to the Campo dei Frari in Venice, and by his overworked imagination. He imagined that

one day it will be different. The luxury homes will be demolished, the Embankment ploughed into the soil. The banks of the Thames will be a series of pleasure gardens, boat houses, jetties and beaches. There will be markets and bandstands, landscaped orchards, a deer park. The river will be lined with restaurants serving fish freshly caught from the sweet waters themselves. Children will swim in the shallows and play in the sand . . .[8]

One of the few occasions when I came across his advocacy of access to water in real life, was twenty-five years ago when Lawrence Halprin described the ponds and fountains he was commissioned to build in Portland, Oregon. Municipal water projects are seldom (unlike commercial water parks) intended for personal contact. Nowadays they are usually the centre-piece of traffic islands or the forecourt decoration of buildings of state. In ancient cities on the tourist route, of course, they are often colonized by children, diving for the coins thrown in by visitors. Halprin was faced with the task of making modern ponds in a modern city, and did so with *use* in mind:

We thought a great deal about the feelings people have about water. There seems to be a deeply felt need to become involved with it. So we worked with the idea of allowing the people in the area – particularly the young people – to participate actively in using the fountains. And this idea influenced the process of design. It meant, for example, that the design could not have railings or, for that matter, any constraining elements that would by implication say 'stay out'. The very nature of the forms and boundaries had to imply 'come in, participate, get involved, please use'. The design had to be permissive and indeterminate, to the extent that we ourselves should not know what would emerge in the participation.[9]

There actually is a parallel between our common enjoyment of public water belonging to no one and shared by all, and our individual need for adequate and safe domestic water. Seas, rivers, lakes and fountains are not private assets, although access to them may be. Rainfall belongs to no one, but its purification and delivery to your kitchen and bathroom are a vital service for which we pay, individually or collectively. Experience showed that for the good of society as a whole, every household should have a supply, however little they paid. Before this lesson was learned, in 1844, Elizabeth Stubbs (see Chapter 1) was fined for taking water from the Preston Water Company's tap without the contract to do so. It is profoundly depressing to read that 150 years later, Rachel and Steve, with their water supply cut off, had to admit that 'We went round to the next door neighbours and filled up the bath with water through their hose and used that for washing, toilet, food and that sort of thing.'[10] They were, theoretically, guilty of an identical offence.

The fact that their plight would have been inconceivable in a British city in the twentieth century until the 1980s is a reminder that the dogmas of market economics have been absorbed in Britain with the overwhelming force of a religious revival. They have changed the language of all of us, whether converts or not, with the result that water-users, just like rail passengers, are now described as 'customers'. One correspondent of the *New Statesman* urges, correctly in my view, that 'We need to recognise that "business studies" as currently constituted amounts to the most successful programme of political propaganda ever undertaken in this country.'[11] The same assumptions have for many years dominated the policies of international agencies like the World

Bank and the International Monetary Fund, in their approach to the water problems of the poor world.

Every such evangelistic wave is followed by a reaction reasserting older values. This is why in the Preface to this book I drew attention to the findings of Richard Titmuss in comparing the blood donor system (based on a sense of solidarity and social responsibility towards other, unknown, members of society) and the commercial market in blood, which he found to be essentially a redistribution from the poor to the rich. I suggested that there are parallels as well as differences between the distribution of these two necessities of life. And in the course of this book I have shown that in the vital social duty of conserving water, the British experience indicates a difference in public attitudes between water supplied as a public good and water supplied as a commercial transaction.

Years earlier Titmuss had argued that 'Social ideas may well be as important in Britain in the next half-century as technical innovation.'[12] Most of that half-century has passed, and all we have encountered in the form of social ideas has been what he condemned as 'the philistine resurrection of economic man in social policy'.[13]

Yet, as this book has shown, a variety of human societies around the world have evolved sophisticated systems of water distribution which combine water conservation with an automatic respect for fairness and reciprocity. We are faced not with a technical problem, but with a crisis of social responsibility.

Notes

1. William Corlett, *Now and Then* (London: Abacus, 1995), p. 190.

2. Jean Robert, *Water is a Commons* (Mexico D. F.: Habitat International Coalition, 1994), p. 112

3. Anna Davin, *Growing Up Poor: Home, School and Street in London* (London: Rivers Oram Press, 1995), p. 64.

4. Jill Ker Conway, *The Road from Coorain* (London: William Heinemann, 1989), p. 32.

5. Georges Simenon, *Je me souviens* (1945), quoted in Patrick Marnham, *The Man Who Wasn't Maigret* (London: Bloomsbury, 1992), p. 19.

6. Douglas Goldring, quoted in Colin Ward: *The Child in the City* (London: Architectural Press, 1979), p. 93.

7. Sean French, in *New Statesman & Society*, 29 March 1996

8. *Ibid.*

9. Lawrence Halprin, 'Annual discourse at the Royal Institute of British Architects', *RIBA Journal*, July 1971.

10. Alicia Herbert and Elaine Kempson, *Water Debt and Disconnection* (London: Policy Studies Institute, 1995), p. 65.

11. Brendan Lambon, 'Propaganda studies', letter in *New Statesman & Society*, 26 January 1996, p. 28.

12. Richard Titmuss, *Essays on 'The Welfare State'* (London: Allen & Unwin, 1958), p. 12.

13. Richard Titmuss, *The Gift Relationship* (London: Allen & Unwin, 1970), p. 14.

APPENDIX

Information and Action

This book is a patchwork quilt of a great variety of sources of information listed in my source notes. I would like readers to explore the issues further, and need to cite a few readily accessible books with different points of view.

The first, relating British and world experience, is:

David Kinnersley, *Coming Clean: The Politics of Water and the Environment* (Harmondsworth: Penguin Books, 1994).

The second, with a global perspective, is:

Robin Clarke, *Water: The International Crisis* (London: Earthscan Publications, 1991).

The third, with a formidable account of large-scale solutions, is:

Fred Pearce, *The Dammed: Rivers, Dams and the Coming World Water Crisis* (London: The Bodley Head, 1992).

Readers with access to central London should know that there is one bookshop with a remarkable range of water literature. This is the IT Bookshop, at 103–105 Southampton Row, London WC1B 4HH, Tel: 0171 436 9761. It issues regular catalogues covering many publishers and several languages, and the quarterly journal *Waterlines*.

Oxfam Publications, 274 Banbury Road, Oxford OX2 7DZ, have a similar wide range of general and specialist literature. So does Earthscan Publications, 120 Pentonville Road, London N1 9JN, which publishes in association with the International Institute of Environment and Development. The charity WaterAid, founded within the water industry when it was publicly owned, produces a quarterly journal for supporters, *Oasis*, from 1 Queen Anne's Gate, London SW1H 9BT.

The most recent study of public attitudes to water conservation and payment for water in England and Wales is *Water Consumption and Charges: Policy Report*, published in July 1996 by the Consumers Association, 2 Marylebone Road, London NW1 4DF.

Index

Adahur, Tarak   78
Adams, Patricia   88
Afghanistan   77
Africa   36, 53, 57, 67, 80, 83, 84, 89, 108, 110, 112, 113
agriculture   21–7, 37, 41–2, 49–51, 65, 76–7, 78–80, 110–13, 117, 126, 135–6
Akasombo Dam   35
Ali, Mohammed   50, 51
Aragon   22, 24, 25
Aral Sea   51, 52, 112
Argentina   68, 69, 96–7, 98–100, 109
Arun project, Nepal   55–6
Aswan High Dam   50–1, 64
Aswan Low Dam   50
Australia   75, 107, 108, 118, 135

Bailey, Brian   2–3
Bali   21, 78
Bangladesh   58, 69, 70, 87, 123, 125
Barak river   70, 124
beaches   117, 120
Beas-Sutlej River   69
Bhutan   69, 70
Birmingham   63, 103–4, 93, 116
Bolivia   68
bottled water   12, 118
Boutros-Ghali, Boutros   51, 124
Brahmaputra River   69–70, 124
Brazil   68

Bretton Woods Conference   53
Britain, British Isles   1, 8, 10, 20, 49, 50, 61, 63, 73, 90, 92–6, 101, 115–21, 126–30, 138
Buber, Martin   66
Budd, William   6
Burkina Faso   113

California   31, 107–8
Cambodia   36
Canada   27, 48
canals   22, 49, 50, 55, 75, 85
Carr, Raymond   21, 23
Chad   113
Chile   68
China   32, 34–6, 52, 70, 76, 123
cholera   2, 5, 6, 12, 118–19
Cinca River   25
Clarke, Robin   21, 34, 35, 64, 68–70, 76–7
Colorado River   68
community control of water   1, 17, 19–21, 24–6, 35, 58–9, 62–3, 75, 78–9, 84–5, 99–100, 109
Cornwall   116
Costa, Joaquin   15, 22, 23, 26
cotton   50, 51, 80, 112, 128
Croall, Jonathan   110
Cruachan Dam   64–5
Crump, Andy   125
Cullis, Adrian   77, 81
Curtis, Donald   79

143

INDEX

dams   35–57 *see also* individual dams
Davidson, Joan   57, 67, 111–13
desalination   67, 110
Devon   116
diarrhoea   131
Didion, Joan   31, 36
disconnection of domestic water   viii, 11–12, 93, 96, 103
disease, waterborne   11, 57, 67, 85–6, 88, 131
Dneiper Dam   47
Dobson, Frank   129
Dogra, Bharat   ix, 70–1, 124–5
Doxiadis, Constantinos   83–4
drinking water   1–3, 6–7, 12, 76, 86–7, 98, 107–9, 119
drought   26, 79–80, 94–5, 135

East Anglia   127
Ecuador   97–8
Egypt   32, 49–51, 64, 112, 123
El Salvador   85
England and Wales
  re-organization of water industry (1974)   8–9, 93, 128
  privatization of water industry (1989)   10–12, 95–6, 121, 138
Ethiopia   64, 81, 113
Euphrates, River   64
European Union   117, 120–1
Evans, George Ewart   3

Fernes, Robert   33
fertilizers   3, 42–3, 51, 112, 120–1
Ford, Henry   40
forests   20, 23, 37, 56, 69, 113
France   96, 130
Franco, Francisco   23, 25–7
French, Sean   136–7

Gandhi, M. K.   59, 123, 125
Ganges, River   69, 70, 124–5
Geddes, Patrick   38, 39, 58–9, 76
Gellner, Ernest   33–4

Germany   53, 83
Ghana   35, 53, 64, 84
Gibraltar   76, 110
global warming   123–5
golf courses   27, 110–11
Granada   23, 133
Grand Coulee Dam   36
Grande, Rio   68
Guinea-Bissau   88
Gummer, John Selwyn   95, 120, 129

Hall, Peter   38–43
Halprin, Lawrence   137
Hardin, Garrett   15, 16, 19–21, 27
Hennessy, Peter   48
Herrington, Paul   101–3, 128–9
Holme, Samuel   5
Hooper, John   27
Hoover Dam   36
Houghton, Sir John   123–4
Humber, River   120
Hungary   107–8
Hurd, Douglas   57
Huxley, Julian   37
'hydraulic societies'   31–3

Illich, Ivan   62, 131
India   47, 52, 53, 55, 57–8, 69–70, 74, 76–7, 79–80, 124–5
Indonesia   35
Intermediate Technology   73–5, 140
International Monetary Fund   53
Ipswich   4, 5, 115
Iraq   33–4, 64, 67
irrigation   21–6, 33–5, 49–50, 51–2, 55, 71, 79, 111, 124–5, 126, 128, 136
Israel   64, 65–6, 76–7, 111

Jackson, Helen   11
Jacobson, Howard   111
Johnston, Tom   48
Jordan, River   65–6

INDEX

Kedung Ombo Dam 25
Kinnersley, David 1, 83–4, 88–9, 117–18
Koenigsberger, Otto 65
Kropotkin, Peter 38–9, 91, 101
Kuwait 67

Labajo, Antonio 27
Lansbury, George 136
Lauca, River 68
Leach, Edmund 33
Lebanon 66
Lenin, V. I. 47
Leval, Gaston 25
Libiszewski, Stephan 63
Libya 52
Lilienthal, David 40
Liverpool 5, 63
London 6, 56, 64, 94, 116, 123, 134–5, 136–7
Lopez de Urralde, Juan 27
Luckin, Bill 107
Luttwak, Edward 127

MacKaye, Benton 39
McRobie, George 73, 74
Mahaweli Dam 34
Malaysia 56–7
Malefakis, Edward 23, 24
Mali 113
Marshall, Peter 18–19
Maura, Antonio 23
Mekong River 36
Mexico 62, 67–8, 76, 118–19
Middle East 9, 51, 53, 64, 67, 76, 111
Middleton, John 11
Miller, Duncan 78
'mini-dams' 38, 58, 71, 77, 79–80
Min River 34–5
Mississippi River 37, 41
Morgan, A. E. 40
Morgan, Harcourt 41, 42
Morgan, Nigel 5

Mulvany, Patrick 81
Mumford, Lewis 39
municipal control of water 3–4, 6, 8, 93, 96–7, 99–100, 103

Narmada Sagae Dam 55
Nasser, Gamal Abdel 50
National Rivers Authority 12, 121, 127
National Water Council 8, 94
Nepal 20, 55–6, 69, 70, 78–9
Nile, River 49–51, 64, 112
Norris, George 40
Northern Ireland 11
Nyerere, Julius 111

Ohio River 37, 41
Organization for Economic Co-operation and Development (OECD) 78, 101–2
Oroville Dam, 31, 36
Orwell, River 4
Oxfam 54, 57, 74, 113, 141

Pacey, Arnold 77, 81
Pakistan 58, 69, 85
Palestine 65, 111
Paraguay 47, 68
Paraná, River 68
parish pump 2–3, 12, 83
Parker, Dennis 61
Patkar, Medha 55
Pearce, Fred 8, 9, 26, 34, 36, 47, 50, 52, 55, 65, 68, 94
Penning-Rowsell, Edmund 61
Pergau Dam 56–7
Peru 92–3, 96, 100–1, 118
Plata, Rio de la 68
pollution 3–5, 59, 86, 88, 115–21
population 16–18, 89, 100, 116, 119
Postel, Sandra 92, 101
Preston, Lancs 5, 12, 138

INDEX

pricing of water, methods of   10, 100–3
Public Health Acts   4

rainwater harvesting   76–8
Ravi, River   69
Reclus, Elisée   38
recycling of water   1, 76–7, 95, 116
regionalism   16–17, 58, 62, 63, 71
reservoirs   35, 40, 49–50, 63
Rio Grande   68
river basins   7, 32, 37–42, 49–50, 52, 62–6, 71
Robert, Jean   62, 63, 70, 91, 98, 102, 118, 119, 121, 133
Rogun Dam   47
Roosevelt, F. D.   40, 43, 44
Roosevelt, T.   39
Royal Commission on Environmental Pollution   121
'run-off'   75–6

Sadat, Anwar   64
Sale, Kirkpatrick   16, 17
salination   32, 51, 52, 68, 79, 125
Sands, Geoff   85
sanitation, non-waterborne   62, 115–16, 118
Sardar Sarovar Dam   53
Sarin, Madhu   89
Satterthwaite, David   108–9
Schumacher, E. F   73–4
Scotland   11, 48–9
sedimentation   26, 50, 79
Selincourt, Kate de   35
Seregeldin, Ismail   66–7
Severn, River   120
sewage treatment   7–8, 59, 62, 115–19
sewage, untreated   4, 7, 86, 120–1
Sewar, River   78–9
Simenon, Georges   135–6
Snell, K. D. M.   20
Snow, Dr John   6

Solomon, Lawrence   88
Somalia   80
South Africa   108
Spain   15, 21–7, 75, 133
Sri Lanka   33–4
Stein, Clarence   39
Stubbs, Elizabeth   5, 12, 138
Sudan   80, 81
Suffolk   1–5
Switzerland   134
Swyngedouw, Erik   97–8
Syria   64

Tagore, Rabindranath   59, 125
Tamini, Abdul Rahman   111
Tanzania   111
Tennessee River   37, 40–1
Tennessee Valley Authority   37–48
Thailand   36, 87, 110–11
Thames, River   6, 7, 116, 137
Three Gorges project   36, 52
Tibet   34–5, 70
Tigris, River   64
Titmuss, Richard   vii–viii, ix, 139
'Tragedy of the Commons'   15–21
Turner, Jean   115
Turner, John   62
Turkey   47, 64
Twain, Mark   61

Uganda   80–1
United Kingdom see Britain
United Nations   65, 98, 108
Uruguay   100
USA   27, 31, 36–44, 47, 48, 50, 67–8, 74, 76, 112, 119, 134
USSR   32–4, 50–1, 112

Valencia   22, 25, 26, 133
Vidal de la Blache, Paul   38
Vietnam   36
Volta, River   35, 53, 64, 65

INDEX

Wales 48, 63 *see also* England and Wales
WaterAid 89, 142
water industry in England and Wales
   reorganization 8–9
   privitization 10–11
water, methods of pricing 91–104
Wijk-Sijbesma, Christine van 85
Wittfogel, Karl 32–4, 44, 49
Wohl, Anthony 6
women and water 82–90
Woodcock, George 79
Woollands, Héctor 99–100
World Bank 53–6, 62, 69, 111

World Health Organization 85–6, 108–9
Worpole, Ken 107–8
Wright, Henry 39

Xanten, Germany 83
Xiao Langdi Dam 36

Yacyreta Dam 68
Yangtse River 52
Yellow River 36

Zambesi River 36